Third Edition

Peter C. Sessler

MBI Publishing Company

This third edition published in 2000 by MBI Publishing Company, 729 Prospect Avenue, PO Box 1, Osceola, WI 54020-0001 USA

©Peter C. Sessler, 1990, 1995, 2000

All rights reserved. With the exception of quoting brief passages for the purposes of review, no part of this publication may be reproduced without prior written permission from the Publisher.

The information in this book is true and complete to the best of our knowledge. All recommendations are made without any guarantee on the part of the author or Publisher, who also disclaim any liability incurred in connection with the use of this data or specific details.

We recognize that some words, model names and designations, for example, mentioned herein are the property of the trademark holder. We use them for identification purposes only. This is not an official publication.

MBI Publishing Company books are also available at discounts in bulk quantity for industrial or sales-promotional use. For details write to Special Sales Manager at Motorbooks International Wholesalers & Distributors, 729 Prospect Avenue, PO Box 1, Osceola, WI 54020-0001 USA.

Library of Congress Cataloging-in-Publication Data
Sessler, Peter C.,
    Mustang red book 1964-1/2to 2000/ Peter C.
      Sessler.-- 3rd ed.
        p.cm.
    Includes index
    ISBN 0-7603-0800-4 (pbk. : alk. paper)
      1. Mustang automobile--History. I. Title.

TL215.M8 S484 2000
629.222'2--dc21

On the front cover: This beautiful example of a 1966 Fastback 2+2 is owned by Bill Baxter. *David Newhardt*

Printed in the United States of America

Contents

	Introduction	4
1	1965 Mustang	6
2	1966 Mustang	13
3	1967 Mustang	17
4	1968 Mustang	22
5	1969 Mustang	27
6	1970 Mustang	35
7	1971 Mustang	41
8	1972 Mustang	48
9	1973 Mustang	52
10	1974 Mustang	56
11	1975 Mustang	60
12	1976 Mustang	64
13	1977 Mustang	68
14	1978 Mustang	73
15	1979 Mustang	78
16	1980 Mustang	83
17	1981 Mustang	88
18	1982 Mustang	92
19	1983 Mustang	96
20	1984 Mustang	100
21	1985 Mustang	105
22	1986 Mustang	109
23	1987 Mustang	112
24	1988 Mustang	116
25	1989 Mustang	118
26	1990 Mustang	120
27	1991 Mustang	122
28	1992 Mustang	124
29	1993 Mustang	126
30	1994 Mustang	129
31	1995 Mustang	133
32	1996 Mustang	136
33	1997 Mustang	139
34	1998 Mustang	142
35	1999 Mustang	145
36	2000 Mustang	148
37	1993 SVT Mustang Cobra	150
38	1994 SVT Mustang Cobra	153
39	1995 SVT Mustang Cobra	157
40	1996 SVT Mustang Cobra	160
41	1997 SVT Mustang Cobra	163
42	1998 SVT Mustang Cobra	165
43	1999 SVT Mustang Cobra	167
44	2000 SVT Mustang Cobra	170
45	1965 Shelby Mustang	172
46	1966 Shelby Mustang	174
47	1967 Shelby Mustang	177
48	1968 Shelby Mustang	180
49	1969-70 Shelby Mustang	183
	Appendix	186

Introduction

The *Mustang Red Book* is designed to help the Mustang enthusiast determine authenticity and originality of any production Mustang and Shelby Mustang built from 1965 through 2000. This updated third edition now includes separate chapters on the 1993–2000 SVT Mustang Cobra. Each chapter covers a single model year of Mustang production and lists production figures, VIN decoding information, engine codes, exterior color codes, interior trim codes, pricing and option information and selected facts.

Every effort has been made to ensure the accuracy of the information presented here. And, it is applicable to the great majority of Mustangs built. However, there are exceptions. The possibility of shortages, substitutions, deletions and the like is always there, which results in a Mustang that doesn't quite fit published specifications. Most of the variations you'll find are relatively minor detail items, not something that changes the basic specifications of the car in question. For example, I doubt you'll find a six-cylinder Shelby.

The most important number in any Mustang is the VIN or Vehicle Identification Number. From 1965 through 1980, this is an 11-digit number that breaks down to model year, assembly plant, body code, engine code and consecutive unit number. Starting with 1981, Mustangs used a 17-digit number which, in addition to the previous information, includes a vehicle manufacturer code and type of passenger restraint system used. The VIN is stamped on various places on the car, sometimes requiring disassembly to see it. For titling and registration purposes, there is one specific area where the manufacturer must affix the vehicle's VIN. For 1965–67 Mustangs, the number was stamped on the driver's side inner fender panel on a notch, between the shock tower and radiator support. In 1968, it was stamped on a plate which was riveted to the edge of the instrument panel, on the passenger side, and was visible through the windshield. From 1969 to the present, the plate was moved over to the driver's side.

The VIN, along with additional information (see appendix), is also stamped on a metal warranty plate and riveted on the rear face of the driver's door on Mustangs built 1965–69. From 1970 to the present, Ford stopped using metal plates, substituting a vehicle certification label.

The additional information consists of body type, color code, interior trim code, axle ratio code, transmission code and DSO (District Special Order), which is an internal Ford code used for production. This warranty plate and label have, unfortunately, been abused by some Mustang owners, and for the wrong reasons.

Many Mustang vendors sell blank VIN and warranty plates and labels, for Mustang owners who want a "new" plate to replace a damaged original or just to install one that reflects the Mustang's true original state, especially if the original door was replaced somewhere along the way. Most vendors require that the present owner verify through a copy of a title or other papers the Mustang's original VIN.

Mustangs also have a "broadcast sheet"—a printout showing the VIN and the options that the car was built with. It is usually located underneath the carpets or sometimes taped around one of the wiring harnesses beneath the dash. An original broadcast sheet is a good way to check originality.

With the exception of 1965–66 (and some 1967) Shelby GT350s, Boss 302s and Boss 429s, you cannot match a specific engine with the Mustang it was originally installed in. Ford did not stamp the block or heads with the car's VIN. However, with the Shelby and Boss Mustangs, knowing that the numbers match increases authenticity.

Ford also attached a stamped plate, usually under one of the coil hold-down bolts, which supplied engine information. Carburetors, too, were tagged.

There is also a riveted underhood plate with additional information such as the consecutive unit number and option codes. Its location varies with the plant that the car was built in and its year of manufacture. Some places to look for it are behind the headlights on the radiator support, near the right-side hood hinge and on the sides of the engine compartment. Every Mustang built has one.

Each chapter lists the published exterior colors and interior trim codes. Don't forget, though, that Ford built and will build Mustangs in colors and with different interior trims than those normally listed. The color section on the warranty plate or label in such cases is usually left blank.

As said before, this book will help you establish authenticity and value. However, in the final analysis, you have to use the information presented here along with other factors, such as common sense, to make a judgment.

Although every effort has been made to make sure that the information contained in the *Mustang Red Book* is correct, I cannot assume any responsibility for any loss arising from use of this book.

Special Thanks To

Paul McLaughlin, Rick Kopec, Eric Binns, Wayne Houghtaling, Kevin Marti and his book, *Mustang by the Numbers 1967–1973*, Monty Montero, Carol Kuperman, SAAC's Shelby American World Registry, SVT Cobra Owners Association, and Ford Motor Company.

Chapter 1

1965 Mustang

Production Figures — Early 1965

65A 2dr Hardtop	92,705
76A Convertible	28,833
Total	121,538

Production Figures — Late 1965

63A 2dr Fastback, standard	71,303
63B 2dr Fastback, luxury	5,776
65A 2dr Hardtop, standard	372,123
65B 2dr Hardtop, luxury	22,232
65C 2dr Hardtop, bench seats	14,905
76A Convertible, standard	65,663
76B Convertible, luxury	5,338
76C Convertible, bench seats	2,111
Total	559,451
Total — Early & Late	680,989

Serial Numbers

5F07F100001
5 — Last digit of model year
F — Assembly plant (F-Dearborn, R-San Jose, T-Metuchen)
07 — Plate code for 2dr Mustang (08-convertible, 09-fastback)
F — Engine code
100001 — Consecutive unit number

Location
Stamped on driver's side inner fender panel, at notch in fender between shock tower and radiator support; warranty plate is riveted on rear face of driver's door.

Engine Codes
U — 170 ci 1V 6 cyl 101 hp (early 1965)
T — 200 ci 1V 6 cyl 120 hp
F — 260 ci 2V V-8 164 hp (early 1965)
C — 289 ci 2V V-8 200 hp
D — 289 ci 4V V-8 210 hp (early 1965)
A — 289 ci 4V V-8 225 hp
K — 289 ci 4V V-8 271 hp high performance

V-8 Distributors

260 ci 164 hp — C40F-12127-A/manual,
C40F-12127-B/automatic
289 ci 200 hp — C5AF-12127-M or CZAF-12127-C or
C5GF-12127-A
289 ci 210 hp —C5AF-12127-M/manual,
C5AF-12127-N/automatic

V-8 Distributors
289 ci 225 hp — C5AF-12127-M/manual,
C5AF-12127-N/automatic
289 ci 271 hp — C30F-12127-D

V-8 Carburetors
260 ci 164 hp — C40F-9510-A/manual, C40F-9510-B/automatic
289 ci 200 hp — C5ZF-9510-A/manual, C5ZF-9510-B or H/automatic
289 ci 210 hp — C4GF-9510-U/manual, C4GF-9510-E or V/automatic
289 ci 225 hp — C5ZF-9510-C or J/manual, C5ZF-9510-D or K/automatic
289 ci 271 hp — C40F-9510-AL, C50F-9510-J, L/manual
C40F-9510-AT, C50F-9510-K, M/automatic

Steering Gear Ratios
HCC-AT — 19.9:1
HCC-AX — 16:1
HCC-AW — 16:1

1965 Mustang Prices

	Retail
Hardtop, 65A	$2,320.96
Convertible, 76A	2,557.64
2+2 Fastback, 63A	2,533.19
200 hp 289 V-8 extra charge over 120 hp 6 cyl	105.63
225 hp 289 V-8 extra charge over 200 hp V-8	52.85
271 hp 289 V-8 without GT Equipment Group	327.92
271 hp 289 V-8 with GT Equipment Group	276.34
Cruise-O-Matic automatic transmission, 6 cyl	175.80
Cruise-O-Matic automatic transmission, 200 & 225 hp V-8s	185.39
4-speed manual transmission, 6 cyl	113.45
4-speed manual transmission, V-8 engines	184.02
Manual front disc brakes, 8 cyl	56.77
Limited slip differential	41.60
Rally-Pac clock/tachometer	69.30
Special Handling Package, 200 & 225 hp V-8s	30.64
GT Equipment Group	165.03
Styled steel wheels, 8 cyl only	119.71
Power brakes	42.29
Power steering	84.47
Power convertible top	52.95
Emergency flashers	19.19
Padded visors, 65A & 76A	5.58
Seatbelts, rear	14.78
Deluxe seatbelts, front (retractable)	7.39
Deluxe seatbelts, front & rear (front retractors)	25.40
Visibility Group	35.83
Accent Group, 65A & 76A	27.11
Accent Group, 63A	13.90
Air conditioner, Ford	277.20

Back-up lamps	10.47
Battery, heavy-duty	7.44
Closed emission system (Calif. type)	5.19
Full-length console	50.41
Console, with air conditioner	31.52
Interior Decor Group	107.08
Full-width seat (bench) with center armrest, 65A & 76A	24.42
Tinted glass with banded windshield	30.25
Windshield only, tinted & banded	21.09
Push-button radio & antenna	57.51
Rocker panel molding, 65A & 76A	15.76
Deluxe steering wheel	31.52
Vinyl roof, 65A	74.19
Wheel covers, knock-off hubs	17.82
MagicAire heater, delete (credit)	(31.52)
Seatbelts, delete (credit)	(10.76)
Tires, 6 cyl extra charge over 6.50x13 4 p.r. BSW	
(5) 6.50x13 4-p.r. WSW	33.30
(5) 6.95x14 4-p.r. BSW	7.36
(5) 6.95x14 4-p.r. WSW	40.67
Tires, 8 cyl extra charge over 6.95x14–p.r. BSW	
(5) 6.95x14 4-p.r. WSW	33.31
(5) 6.95x14 4-p.r. BSW nylon	15.67*
(5) 6.95x14 4-p.r. WSW nylon	48.69*
(5) 6.95x14 4-p.r. Dual Red Band nylon (std. 271 hp)	48.97

* NC with 271 hp 289 V-8

Dealer-added Accessories

Door edge guards	$ 2.70
Rocker panel molding (set)	19.10
Deluxe with spinner 13 in wheel covers	28.95
Deluxe with spinner 14 in wheel covers	28.95
Simulated wire 13 in wheel covers	58.35
Simulated wire 14 in wheel covers	58.35
Luggage rack	35.00
Tonneau cover, white	52.70
Tonneau cover, black	52.70
Lefthand spotlight	29.95
Vanity mirror	1.95
License plate frame	4.50
Fire extinguisher	33.70
Compass	7.95
AM radio	53.50
Rear seat speaker	11.95
Studio Sonic Sound System (Reverb)	22.95
Round (cone-shaped) outside mirror	3.95
Lefthand remote mirror	2.25
Universal (flat) outside mirror	12.75
Matching righthand mirror	6.75
Inside day-night mirror	4.95
Back-up lights	10.40
Power brakes	47.00

Glovebox lock	2.49
Remote-control trunk release	6.95
Windshield washers	14.50
Rally-Pac, 6 cyl	75.95
Rally-Pac, 8 cyl	75.95

1965 (early) Exterior Colors — Code

Raven Black	A
Pagoda Green	B
Dynasty Green	D
Guardsman Blue	F
Caspian Blue	H
Rangoon Red	J
Poppy Red	3
Silversmoke Gray	K
Wimbledon White	M
Prairie Bronze	P
Cascade Green	S
Sunlight Yellow	V
Vintage Burgundy	X
Skylight Blue	Y
Chantilly Beige	Z
Twilight Turquoise	5
Phoenician Yellow	7

1965 (early) Interior Trim — Code

Parchment vinyl w/blue	42
Parchment vinyl w/red	45
Parchment vinyl w/black	46
Parchment vinyl w/Ivy Gold	48
White vinyl w/Palomino	49
Black vinyl & cloth	56
Blue vinyl w/blue	82
Red vinyl w/red	85
Black vinyl w/black	86
Palomino vinyl w/Palomino	89

1965 (late) Exterior Colors — Code

Raven Black	A
Midnight Turquoise	B
Honey Gold	C
Dynasty Green	D
Caspian Blue	H
Champagne Beige	I
Rangoon Red	J
Poppy Red	3
Silversmoke Gray	K
Wimbledon White	M
Tropical Turquoise	O
Prairie Bronze	P
Ivy Green	R
Sunlight Yellow	V
Vintage Burgundy	X
Silver Blue	Y
Springtime Yellow	8

1965 (late) Interior Trim — Code

Blue vinyl w/blue	22
Red vinyl w/red	25
Black vinyl w/black	26
Aqua	27
Ivy Gold w/gold	28
Palomino vinyl w/Palomino	29
Parchment w/blue	D2
Parchment w/Burgundy	D3
Parchment w/red	D5
Parchment w/black	D6
Parchment w/Aqua	D7
Parchment w/Ivy Gold	D8
Parchment w/Palomino	D9
Blue & White, luxury	62
Red, luxury	65
Black, luxury	66
Aqua & white, luxury	67
Ivy Gold & white, luxury	68
Palomino, luxury	69
Parchment w/blue, luxury	F2
Parchment w/Burgundy	F3
Parchment w/Emberglo	F4

1965 (late) Interior Trim	**Code**
Parchment w/red, luxury	F5
Parchment w/black, luxury	F6
Parchment w/aqua, luxury	F7
Parchment w/Ivy Gold, luxury	F8
Parchment w/Palomino, luxury	F9
Blue, bench	32
Red, bench	35
Black, bench	36
Palomino, bench	39
Black fabric & vinyl	76
Palomino fabric & vinyl Parchment	79
Parchment w/blue, bench	C2
Parchment w/Burgundy, bench	C3
Parchment w/Emberglo, bench	C4
Parchment w/black, bench	C6
Parchment w/Aqua, bench	C7
Parchment w/Ivy Gold, bench	C8
Parchment w/Palomino	C9

1965 Mustang Facts

Mustangs built between March 1964 and August 17, 1964 were known as early 1965 Mustangs. There were no 1964½ Mustangs. Those built after August 17, 1964 were known as late 1965s. A quick way to identify an early versus a late 1965 is by the electrical system. All early 1965s came with generators; late 1965s came with alternators. Generator-equipped cars came with a GEN warning light on the instrument panel; alternator-equipped Mustangs came with an ALT light. There were a multitude of other differences as well. Some worth noting are: early cars got a stationary passenger seat, a smaller handle on the automatic transmission and larger horns.

There was considerable difference between early and late 1965 Mustang engines. The base 101 hp 170 ci six-cylinder, the 164 hp 260 ci V-8 and the 210 hp 289 ci V-8 were all replaced by a 120 hp 200 ci six-cylinder, a 200 hp 289 ci V-8 and a 225 hp 289 ci V-8, respectively.

Starting in June 1964, the famous 271 hp 289 V-8 was available, and only with a four-speed manual transmission. It came with the Special Handling Package (stiffer springs, shocks and front stabilizer bar) and the fourteen-inch Red Band tires. Axle ratio choice was limited to 3.89:1 or 4.11:1. Incidentally, the Special Handling Package was available only on the 289 V-8 powered Mustangs. All Mustangs with the 271 hp 289 also came with the larger nine-inch rear axle ring gear. All other Mustangs got the smaller eight-inch rear. Only 7,273 1965 Mustangs were equipped with this engine.

The third Mustang body style, the fastback, known as the 2+2, was introduced in September 1964.

From March 1965, the Interior Decor Group, more commonly known as the Pony Interior because of the embossed ponies on the seats, became available. The letter "B" indicates this option on the body code on the warranty plate. Thus 65B, 63B and 76B all indicate the luxury interior, while the letter "A" indicates the standard interior. In addition, the Interior Decor Group option also included the five-dial instrument bezel which replaced the standard bezel.

The GT Equipment Group, available on all Mustang body styles, was available from April 1965 and only on the 225 hp and 271 hp 289 V-8 equipped Mustangs. The package consisted of manual front disc brakes, the Special Handling Package, quick ratio steering, chrome exhaust trumpets, rocker panel stripes, GT emblems and grille-mounted foglamps. In the interior, the standard instrument bezel was replaced with a five-dial version. As true GTs were built from February 1965 through August 1965, these Mustangs should have date codes from P to V on the warranty plate.

Only 15,079 1965 Mustangs came with the GT Equipment Group option.

The rarest 1965 Mustangs were the Indianapolis 500 Pace Car convertibles for the 1964 race. Two were used in the actual race and 35 others were given to race dignitaries. All were painted Pace Car White (which is not the same color as Wimbledon White). Additionally, 185 Indianapolis Pace Car replica hardtops were built and given to dealers. All were painted Pace Car White and equipped with the 260 ci V-8 and automatic transmission.

Two versions of the Rally-Pac clock/tachometer unit which mounted on the steering wheel were offered. The low-profile version was made to accommodate the five-dial instrument bezel that came with the Interior Decor Group and GT Equipment Group options.

Six-cylinder engines were painted red; V-8 engines were painted black with gold valve covers and air cleaners.

The Mustang lettering on the front fenders was 4⅜ inches long on early Mustangs; on later 1965s, length was increased to five inches.

Standard tires were 6.50x13 blackwalls on four- or five-lug 13x4.5 inch steel rims. Optional were whitewalls and a larger 7.00x13 size. The thirteen-inch rims were standard equipment on both six-cylinder and V-8 Mustangs, but V-8 Mustangs got five-lug rims. Optionally available were 14x4.5 inch rims with 6.50x14 tires in either four- or five-lug patterns, depending on engine. Standard with the Special Handling Package were 6.50x14 tires on five-lug 14x5.0 rims. Optional with the Special Handling Package were 15x5 inch wheels with 5.90x15 tires.

Late 1965 six-cylinder Mustangs got the same standard thirteen-inch wheels/tires, however the optional fourteen-inch size was upgraded to a 6.95x14. V-8 equipped Mustangs came with five-lug 14x5 inch wheels with the 6.95x14 tires. Standard with the Special Handling Package were 6.95x14 tires, while tires on the Hi-Po (271 hp) 289 powered Mustangs were Dual Red Line 6.95x14s. The fifteen-inch wheels/tires were dropped.

There were exhaust system variations on the 1965 Mustang. Six- and eight-cylinder single exhaust Mustangs came with a single inlet/outlet transverse-mounted muffler sandwiched between the gas tank and axle. Dual exhaust Mustangs used a transverse muffler with twin inlets/outlets plus two additional mufflers, one on each side, in front of the transverse unit. In July 1964, the factory modified the system by deleting the transverse muffler on the dual exhaust system and substituting two resonators, one on each tailpipe.

1965 2+2

1965 convertible

1965 2+2 GT

Chapter 2

1966 Mustang

Production Figures

63A 2dr Fastback, standard	27,809	65C 2dr Hardtop, bench seats	21,397
63B 2dr Fastback, luxury	7,889	76A Convertible	56,409
65A 2dr Hardtop, standard	422,416	76B Convertible, luxury	12,520
65B 2dr Hardtop, luxury	55,938	76C Convertible, bench seats	3,190
		Total	607,568

Serial Numbers
6F07C100001
6 — Last digit of model year
F — Assembly plant (F-Dearborn, R-San Jose, T-Metuchen)
07 — Plate code for 2dr Mustang (08-convertible, 09-fastback)
C — Engine code
100001 — Consecutive unit number

Location
Stamped on driver's side inner fender panel, at notch between shock tower and radiator support; warranty plate is riveted on rear face of driver's door.

Engine Codes
T — 200 ci 1V 6 cyl 120 hp
C — 289 ci 2V V-8 200 hp
A — 289 ci 4V V-8 225 hp
K — 289 ci 4V V-8 271 hp

V-8 Distributors
289 ci 200 hp — C5GF-12127-A
289 ci 225 hp — C5AF-12127-M/manual, C5AF-12127-N/automatic
289 ci 271 hp — C5OF-12127-E

V-8 Carburetors
289 ci 200 hp — C6AF-9510-A/manual, C6AF-9510-B/automatic
289 ci 225 hp — C6ZF-9510-A,D/manual, C6ZF-9510-B,E/automatic
289 ci 271 hp — C6ZF-9510-C/manual, C6ZF-9510-F/automatic

Steering Gear Ratios
HCC-AT — 19.9:1
HCC-AX — 16:1
HCC-AW — 16:1

1966 Mustang Prices

	Retail
2dr Hardtop, 65A	$2,416.18
Convertible, 76A	2,652.86
2+2 Fastback, 63A	2,607.07
200 hp 289 ci V-8 extra charge over 6 cyl	105.63
225 hp 289 ci V-8 extra charge over 200 hp 289	52.85
271 hp 289 ci with GT Equipment Group	276.34
271 hp 289 ci without GT Equipment Group	327.92
Cruise-O-Matic automatic transmission, 6 cyl	175.80
Cruise-O-Matic automatic transmission, 200 & 225 hp V-8s	185.39
Cruise-O-Matic automatic transmission, 271 hp V-8	216.27
4-speed manual transmission, 6 cyl	113.45
4-speed manual transmission, 8 cyl	184.02
Power brakes	42.29
Power steering	84.47
Power convertible top	52.95
Air conditioner, Ford	310.90
AM Radio-Stereosonic tape system (requires radio)	128.29
Front seat, full with arm rest, 65A & 76A	24.42
Luggage rack, rear deck lid, 65A & 76A	32.44
Radio and antenna	57.51
Accent stripe, less rear quarter ornamentation	13.90
Full-length console	50.41
Console, with air conditioner	31.52
Deluxe steering wheel	32.20
Interior Decor Group	94.13
Vinyl roof, 65A	74.36
Wire wheel covers	58.24
Wheel covers, knock-off hubs	19.48
Closed crankcase emissions system	5.19
Exhaust emission control system (NA 271 hp)	45.45
MagicAire heater, delete option	(31.52)
Front disc brakes, 8 cyl (NA with power brakes)	56.77
Limited slip differential	41.60
Rally-Pac clock/tachometer	69.30
Special Handling Package, 200 & 225 hp V-8s	30.64
GT Equipment Group, 225 & 271 hp V-8s	152.20
Styled steel wheels, 14 in, 8 cyl only	93.84
Heavy-duty battery, 55 amp	7.44
Electric windshield wipers, 2-speed	12.95
Deluxe Seatbelts, front & rear (front retractors) and warning light	14.53
Visibility Group (remote mirror, day/nite mirror & 2-speed wipers)	29.81
Tinted glass with banded windshield	30.25
Tinted glass with windshield only	21.09
Optional tires (except with 271 hp 289) extra charge for:	
(5) 6.95x14 4-p.r. WSW	33.31
(5) 6.95x14 4-p.r. BSW nylon (NC with 271 hp)	15.67
(5) 6.95x14 4-p.r. WSW nylon (NC with 271 hp)	48.89
(5) 6.95x14 4-p.r. Dual Red Band nylon	48.97

1966 Exterior Colors

Color	Code
Raven Black	A
Arcadian Blue	F
Sahara Beige	H
Nightmist Blue	K
Wimbledon White	M
Antique Bronze	P
Brittany Blue	Q
Ivy Green Metallic	R
Candyapple Red	T
Tahoe Turquoise	U
Emberglo	V
Vintage Burgundy	X
Silver Blue	Y
Sauterne Gold	Z
Silver Frost	4
Signalflare Red	5
Springtime Yellow	8
Dark Moss Green (late 1966)	Y7

Additional available colors were Medium Palomino Metallic, Medium Silver Metallic, Maroon Metallic, Silver Blue Metallic and Light Beige.

1966 Interior Trim

Trim	Code
Blue w/blue	22
Dark red w/red	25
Black w/black	26
Aqua w/aqua	27
Parchment w/blue	D2
Parchment w/burgundy	D3
Parchment w/Emberglo	D4
Parchment w/black	D6
Parchment w/aqua	D7
Parchment w/Ivy Gold	D8
Parchment w/Palomino	D9
Blue & white, luxury	62
Emberglo & Parchment, luxury	64
Red, luxury	65
Black, luxury	66
Aqua & white, luxury	67
Ivy Gold & white, luxury	68
Parchment w/blue, luxury	F2
Parchment w/burgundy, luxury	F3
Parchment w/Emberglo, luxury	F4
Parchment w/black, luxury	F6
Parchment w/aqua, luxury	F7
Parchment w/Ivy Gold, luxury	F8
Parchment w/Palomino, luxury	F9
Blue, bench	32
Red, bench	35
Black, bench	36
Parchment w/blue, bench	C2
Parchment w/burgundy, bench	C3
Parchment w/Ivy Gold, bench	C8
Parchment w/Palomino, bench	C9

1966 Mustang Facts

The 1966 Mustangs were only slightly restyled. The most noticeable change was the floating horse in the front grille. Other changes included a redesigned gas cap, bright hood lip molding on all models, standard rocker panel moldings (except on the 2+2), standard back-up lights, redesigned side chrome spires (deleted on the GTs and those with accent pinstripes) and revised styled steel wheels. A chrome trim ring was used on the wheel, which now had only the center section chromed.

In the interior, different upholstery patterns and colors were used but the most noticeable change was the use of the five-dial instrument cluster. Padded visors were now standard equipment.

All Mustangs now came with 14x4.5 inch wheels—four lugs on six-cylinder models, five lugs with the V-8s. The standard wheel cover was redesigned. 6.95x14 was the only tire size available, with the 271 hp 289 Hi-Po engine getting premium Dual Red Band nylon tires. Whitewalls were optional.

The standard engine continued to be the 200 ci six-cylinder with the three-speed manual. The Cruise-O-Matic automatic transmission was now optional with the 271 hp "K" 289 V-8. Fewer Mustangs were equipped with this engine in 1966, only 5,469.

The GT Equipment Group still used the horizontal and vertical grille bars found on the 1965s, but did use a redesigned gas cap. The popularity of the GT option increased as 25,517 Mustangs were so equipped.

T-5 was the designation used on Mustangs exported to Germany. All Mustang emblems and names were removed as the Mustang name was used by another manufacturer there. T-5 emblems were used on both front fenders.

The Sprint 200 Option Group was available only with the 200 ci six-cylinder engine. The package, available on all three Mustang bodies, featured wire wheel covers, pinstripes, center console and a chrome air cleaner with a Sprint 200 decal. Most Sprint 200s were hardtops.

All Mustang engines were painted Ford blue.

The Mustang still used the same underdash air conditioning system. The unit's panel was painted black in 1966 and the air outlets did not have an inner trim ring.

The standard wheel cover was a slotted type with a Mustang emblem in the center.

1966 hardtop GT

Chapter 3

1967 Mustang

Production Figures

65A 2dr Hardtop	325,853	76A Convertible	38,751
65B 2dr Hardtop, luxury	22,228	76B Convertible, luxury	4,848
65C 2dr Hardtop, bench seats	8,190	76C Convertible, bench seats	1,209
63A 2dr Fastback	53,651	Total	472,121
63B 2dr Fastback, luxury	17,391		

Serial Numbers
7R01C100001
7 — Last digit of model year
R — Assembly plant (F-Dearborn, R-San Jose, T-Metuchen)
01 — Plate code for 2dr hardtop (02-fastback, 03-convertible)
C — Engine code
100001 — Consecutive unit number

Location
Stamped on driver's side inner fender panel, at notch between shock tower and radiator support; warranty plate is riveted on rear face of driver's door.

Engine Codes
U — 200 ci 1V 6 cyl 120 hp
C — 289 ci 2V V-8 200 hp
A — 289 ci 4V V-8 225 hp
K — 289 ci 4V V-8 271 hp
S — 390 ci 4V V-8 320 hp

V-8 Distributors
289 ci 200 hp — C7OF-12127-A, B, D or E
289 ci 225 hp — C5AF-12127-M/manual, C5AF-12127-N/automatic
289 ci 271 hp — C50F-12127-E
390 ci 320 hp — C7AF-12127-U/non-Thermactor, -F/with Thermactor

V-8 Carburetors
289 ci 200 hp — C7DF-9510-E, G/manual, C7DF-9510-F, H, N, V/automatic
289 ci 225 hp — C7DF-9510-L, C/manual, C7DF-9510-M, D/automatic
289 ci 271 hp — C6ZF-9510-C/manual, C6ZF-9510-C/automatic
390 ci 320 hp — C7OF-9510-A, C (Holley R-3795)/manual, C7OF-9510-B, D, (Holley R-3796)/automatic

Steering Gear Ratios
SMB-A — 19.9:1
SMB-B — 16:1
SMB-C — 19.9:1
SMB-D — 19.9:1
SMB-E — 16:1

1967 Mustang Prices	Retail
2dr Hardtop, 65A	$2,461.46
Convertible, 76A	2,698.14
2+2 Fastback, 63A	2,592.17
289 ci 200 hp V-8	105.63
289 ci 225 hp V-8	158.48
289 ci 271 hp V-8 (with GT Equipment Group only)	433.55
390 ci 320 hp V-8	263.71
Cruise-O-Matic automatic transmission, 6 cyl	188.18
Cruise-O-Matic automatic transmission, 200 or 225 hp V-8	197.89
Cruise-O-Matic automatic transmission, 271 or 320 hp	220.17
4-speed manual transmission, 200 & 225 hp V-8	184.02
4-speed manual transmission, 271 & 320 hp V-8	233.18
Heavy-duty 3-speed manual, required with 320 hp V-8	79.20
Power front disc brakes	64.77
Power steering	84.47
Power convertible top	52.95
GT Equipment Group (with V-8s only)	205.05
Limited slip differential	41.60
Styled steel wheels (2+2 only)	93.84
Styled steel wheels, all others	115.11
Competition handling package (with GT Equip. only)	388.53
Tinted windows and windshield	30.25
Convenience Control Panel	39.50
Fingertip speed control (requires V-8 & Cruise-O-Matic)	71.30
Remote control outside mirror (std. 2+2)	9.58
Convertible safety glass rear window	32.44
SelectAire air conditioner	356.09
AM push-button radio	57.51
AM/FM push-button radio	133.65
Stereosonic tape system (AM radio required)	128.49
2+2 folding rear seat and access door (Sport Deck option)	64.77
Full-width front seat (NA 2+2)	24.42
Tilt-away steering wheel	59.93
Rear deck luggage rack (2+2)	32.44
Comfortweave vinyl trim (NA convertible)	24.53
Center console (requires radio)	50.41
Deluxe steering wheel	31.52
Exterior decor group	38.86
Lower back panel grille	19.48
Interior Decor Group (convertible)	94.36
Interior Decor Group (all others)	108.06
Two-tone paint (lower back grille)	12.95
Accent paint stripe	13.90
Vinyl-covered roof (hardtop)	74.36
Wheel covers (std. 2+2)	21.34

Wire wheel covers (2+2)	58.24
Wire wheel covers (all others)	79.51
Wide Oval Sports tires (V-8 required)	62.35
Whitewall tire option (typical)	33.31
Rocker panel molding (std. 2+2)	15.59
MagicAire heater (delete option)	(31.52)

1967 Exterior Colors

Color	Code
Raven Black	A
Frost Turquoise	B
Acapulco Blue	D
Arcadian Blue	F
Diamond Green	H
Lime Gold	I
Nightmist Blue	K
Wimbledon White	M
Diamond Blue	N
Brittany Blue	Q
Dusk Rose	S
Candyapple Red	T
Burnt Amber	V
Clearwater Aqua	W
Vintage Burgundy	X
Dark moss Green	Y
Sauterne Gold	Z
Silver Frost	4
Pebble Beige	6
Springtime Yellow	8

Additional special 1967 colors were Playboy Pink, Anniversary Gold, Columbine Blue, Aspen Gold, Blue Bonnet, Timberline Green, Lavender and Bright Red.

1967 Interior Trim

Trim	Code
Black	2A
Blue	2B
Red	2D
Saddle	2F
Ivy Gold	2G
Aqua	2K
Parchment	2U
Black, luxury	6A
Blue, luxury	6B
Red, luxury	6D
Saddle, luxury	6F
Ivy Gold, luxury	6G
Aqua, luxury	6K
Parchment, luxury	6U
Black, bench seat	4A
Parchment, bench seat	4U
Black comfortweave	7A
Parchment comfortweave	7U
Black comfortweave, luxury	5A
Parchment comfortweave, luxury	5U

Convertible Top Colors
Black or White

1967 Mustang Facts

The Mustang was completely redesigned for 1967. It was longer and wider, but it did have the same 108 inch wheelbase of its predecessor. The grille opening was enlarged for a decidedly meaner look, while the rear taillight panel was concave. The fastback became a full fastback, and simulated rear quarter panel scoops were used on all three body styles.

The front suspension was widened and redesigned which resulted in a better ride. The optional front disc brakes came with power assist, and the 1967 Mustang was the first to use a dual hydraulic brake system.

Engine selection remained the same as in 1966, but with one major exception. The 390 ci big block, rated at 320 hp, brought serious performance to the Mustang. The 390 used cast iron intake and exhaust manifolds. Carburetion was a single 600 cfm Holley four-barrel. All 390s installed in the Mustang used a dual exhaust system. A total of 28,800 Mustangs were equipped with the 390. The previous hotshot, the 271 hp 289 V-8, saw its popularity wane, as only 472 were built. The GT Equipment Group option was mandatory with the 271 hp 289.

The premier performance Mustang was still the GT. The GT Equipment Group consisted of the grille-mounted foglamps, power front disc brakes, dual exhausts with chrome quad outlets (excluded on the 200 hp 289), F70x14 tires, GT gas cap, the handling package, rocker panel stripes, and GT or GTA (for automatic transmission equipped GTs) emblems. A total of 24,079 Mustangs were equipped with the GT option.

Available only with the GT Equipment Group was the Competition Handling Package, consisting of firmer suspension components, limited slip rear axle and fifteen-inch wheels with wire wheel covers.

The interior was also redesigned. Most noticeable was the new dash, which did away with the Rally-Pac and also featured integral air conditioning. The Tilt-away steering wheel was a new option, as was the fold-down rear seat on the fastback. Other new options included cruise control, a folding glass rear window on the convertibles and an Exterior Decor Group which included a hood with rear-facing louvers that housed turn signal indicators, wheelwell moldings and a pop-open gas cap.

The Interior Decor Group option did not include the galloping Pony inserts.

The Convenience Control Panel housed four warning lights and was located on the dash above the radio (without A/C). With A/C the panel was available only in conjunction with the console. Lights in this case were mounted on either side of the storage compartment. The four lights were: parking brake warning light, door ajar, seatbelt reminder and low fuel.

Standard wheel cover was a 10½ inch hubcap or a 21-spoke wheel cover.

The optional styled steel wheels were wider, necessitating a wider trim ring, and used a blue center cap.

The Dagenham four-speed manual transmission, previously used with the 200 ci six-cylinder, was no longer available.

As with previous Mustangs, there was a long list of dealer-installed options which included the Cobra kits. With these kits, the Mustang enthusiast could improve performance from mild to wild.

Vinyl roof colors were limited to two: black or parchment.

1967 Mustangs used polyethylene-filled ball joints. Manual steering ratio was reduced from a super slow 27:1 to an almost super slow 25.3:1. Power steering ratio was 20.3:1.

1967 convertible

1967 Fastback GT

Chapter 4

1968 Mustang

Production Figures

63A 2dr Fastback	33,585	65C 2dr Hardtop, bench seats	6,113
63B 2dr Fastback Deluxe	7,661	65D 2dr Hardtop Deluxe, bench seats	853
63C 2dr Fastback, bench seats	1,079	76A Convertible	22,037
63D 2dr Fastback Deluxe, bench seats	256	76B Convertible Deluxe	3,339
65A 2dr Hardtop	233,472	Total	317,404
65B 2dr Hardtop Deluxe	9,009		

Serial Numbers

8R01J100001
8 — Last digit of model year
R — Assembly plant (F-Dearborn, R-San Jose, T-Metuchen)
01 — Plate code for 2dr hardtop (02-fastback, 03-convertible)
J — Engine code
100001 — Consecutive unit number

Location

Stamped on plate riveted on passenger's side of instrument panel, visible through the windshield; also stamped on left inner fender; warranty plate is riveted on rear face of driver's door.

Engine Codes
T — 200 ci 1V 6 cyl 120 hp
C — 289 ci 2V V-8 195 hp
F — 302 ci 2V V-8 210 hp
J — 302 ci 4V V-8 230 hp
S — 390 ci 4V V-8 325 hp
W — 427 ci 4V V-8 390 hp
R — 428 ci 4V V-8 335 hp (Cobra Jet)

V-8 Distributors

289 ci 195 hp — C8TF-12127-F/manual, C8OF-12127-C/automatic
302 ci 210 hp — C8AF-12127-E/manual, C8OF-12127-C/automatic
302 ci 230 hp — C8ZF-12127-A/manual, C8ZF-12127-D/automatic
390 ci 325 hp — C7OF-12127-H
427 ci 390 hp — C7OF-12127-F
428 ci 335 hp — C8OF-12127-H/manual, C8OF-12127-J/automatic

V-8 Carburetors
289 ci 195 hp — C8AF-9510-AF/manual,
C8OF-9510-S/automatic
302 ci 210 hp — C8AF-9510-AK/manual,
C8AF-9510-AL/automatic
302 ci 230 hp — C8ZF-9510-A, C or D/manual,
C8ZF-9510-B or D/automatic
390 ci 325 hp — C8OF-9510-C(Holley R-3795)/manual,
C8OF-9510-D(Holley R-3796)/automatic
427 ci 390 hp — C8AF-9510-AD(Holley R-4088)
428 ci 335 hp — C8OF-9510-AA(Holley R-4168)/manual,
C8OF-9510-AB(Holley R-4174)/automatic

Steering Gear Ratios
SMB-D — 19.9:1
SMB-F — 19.9:1
SMB-K — 16:1

1968 Mustang Prices

	Retail
2door Hardtop, 63A	$2,578.60
Convertible	2,814.22
2+2 Fastback	2,689.26
289 ci 195 hp V-8	105.63
302 ci 230 hp V-8	171.77
390 ci 325 hp V-8	263.71
427 ci 390 hp V-8	622.00
428 ci 335 hp V-8*	434.00
SelectShift Cruise-O-Matic, 6 cyl	191.12
SelectShift Cruise-O-Matic, 195/230 hp V-8s	200.85
SelectShift Cruise-O-Matic, 325 hp V-8	233.17
4-speed manual transmission, 195/230 hp V-8s	184.02
4-speed manual transmission, 325 hp V-8	233.18
Power front disc brakes, V-8s (required with 325hp V-8 on GT Equipment Group)	64.77
Power steering	84.47
Power convertible top	52.95
Convertible glass backlite	38.86
GT Equipment Group, 230/325 hp V-8 (NA with Sports Trim Group or optional wheel covers)	146.71
Tachometer (V-8s only)	54.45
Limited slip differential	41.60
Tinted windows and windshield	30.25
Convenience Group (console required with SelectAire)	32.44
Fingertip speed control (with V-8 and SelectShift)	73.83
Remote control outside mirror, lefthand side	9.58
SelectAire air conditioner	360.30
Push-button radio (AM)	61.40
AM/FM stereo radio	181.39
Stereosonic tape system (AM radio required)	133.86
Sport Deck rear seat (2+2 only)	64.77
Full-width front seat (hardtop & 2+2, NA console)	32.44
Tilt-away steering wheel	66.14

Center console (radio required)	53.71
Interior Decor Group (convertible & full-width front seat)	110.16
Interior Decor Group (all others without full-width seat)	123.86
Two-tone hood paint	19.48
Accent paint stripe	13.90
Vinyl-covered roof (hardtop only)	74.36
Wheel covers (NA with GT or V-8 Sports Trim Groups)	21.34
Deluxe wheel covers (NA with GT or V-8 Sports Trim Groups)	34.33
Wide Oval tire option (V-8s only)	78.53
Whitewall tire option	33.31

*Available after April 1, 1968

1968 Exterior Colors

Color	Code
Raven Black	A
Royal Maroon	B
Acapulco Blue	D
Gulfstream Aqua	F
Lime Gold	I
Wimbledon White	M
Diamond Blue	N
Seafoam Green	O
Brittany Blue	Q
Highland Green	R
Candyapple Red	T
Tahoe Turquoise	U
Meadowlark Yellow	W
Presidential Blue	X
Sunlit Gold	Y
Pebble Beige	6

1968 Interior Trim

Trim	Code*
Black vinyl	2A(6A)
Blue vinyl	2B(6B)
Dark red vinyl	2D(6D)
Saddle vinyl	2F(6F)
Ivy Gold vinyl	2G(6G)
Aqua vinyl	2K(6K)
Parchment vinyl	2U(6U)
Nugget Gold vinyl	2Y(6Y)
Black comfortweave, bench	8A(9A)
Blue comfortweave, bench	8B(9B)
Dark red comfortweave, bench	8D(9D)
Parchment comfortweave, bench	8U(9U)
Black comfortweave	7A(5A)
Blue comfortweave	7B(5B)
Dark red comfortweave	7D(5D)
Parchment comfortweave	8U(5U)

*Parentheses indicates with Decor Group

Convertible Top Colors
Black or White

1968 Mustang Facts

The 289 ci 225 hp V-8 was replaced by a 302 ci rated at 230 hp. Increased displacement was achieved by increasing the stroke on the 289 from 2.87 inches to 3.00 inches. The two-barrel carburetor 289 (rated at 195 hp for 1968) was replaced mid-year by a two-barrel 302 rated at 210 hp. Thus both 289- and 302-powered Mustangs were available in 1968.

The base engine, the 200 ci six-cylinder, remained unchanged while the 390 ci V-8 was rated at 325 hp, an increase of five hp. Top engine option was a Low Riser version of Ford's famous 427 ci V-8. Featuring a Holley 600 cfm carburetor and available only with an automatic transmission, it was rated at 390 hp. A rare and expensive option, it was deleted from the option list in December 1967. Look for the letter "W" in the VIN for engine code.

A total of 11,475 1968 Mustangs were equipped with the 390 ci V-8.

On April 1, 1968 Ford introduced a special version of the 428 ci V-8 for use in the Mustang, known as the 428 Cobra Jet. The 428 Cobra Jet was basically a production 428 fitted with 427 Low Riser heads, but with a host of improvements. Rated at 335 hp, it actually produced more than 400. Available only with the GT Equipment Group, 428CJ Mustangs also came with functional Ram Air hood scoop, power front disc brakes and staggered rear shocks for four-speed transmission cars. The C-6 three-speed automatic was also available. A total of 2,253 fastbacks and 564 hardtops were built. An unknown small number of convertibles were built as well.

Fifty pre-production 428CJ powered fastbacks, all painted Wimbledon White, were released before regular Cobra Jet production began. These were sold primarily to racers. They differed from the production versions in several ways. They were all regular Mustangs without the GT option, the 428CJ engine came with an aluminum intake manifold and the shock towers did not have additional bracing. On the window sticker, the option was referred to as the Cobra Jet Program. The 428 engine listed for $507.40.

Goodyear Polyglas F70x14 tires made their debut on the 428CJ Mustangs.

The horizontal grille bars were deleted on 1968 Mustangs and on the GT Mustangs as well. Due to governmental regulations, all Mustangs came with front and rear quarter panel reflectors. Mustangs built before February 15, 1968 came with a rectangular rear reflector while those built after had a bolt-on reflector with oval chrome trim. All 1968 Mustangs came with chrome rocker panel moldings. The simulated side scoops of 1967 were replaced by a vertical ornament.

Two-tone painted louvered hoods were optional on all Mustangs.

The GT Equipment Group was still available on the 230/325/390 hp Mustangs. Differences from the previous year were the new 14-inch styled steel wheels (chromed or painted argent) with GT hubcaps, a new pop-open GT gas cap, new side "C" stripes and new quarter panel GT emblems. There was no separate GTA designation to differentiate automatic-transmission-equipped GTs. GT production dropped to 17,458 in 1968.

The Sprint option package on six-cylinder Mustangs included GT side stripes, pop-open gas cap and full wheel covers. The V-8 Sprint option added GT foglamps and the styled steel wheels with the Wide Oval tires.

The California Special GT/CS was a special trim package available on hardtops in California only. It used a Shelby rear deck lid with integral spoiler and sequential taillights and Shelby non-functional side scoops. The Mustang ornament was deleted from the blacked-out grille opening. Special side stripes, styled steel wheels and Lucas or Marchal foglamps completed the package. About 5,000 were built.

Similar to the California Special was the High Country Special, this time sold only by Colorado dealers. Identical to the GT/CS, with the exception of the High Country Special decal taking the place of the GT/CS identification on the side scoops, the High Country Specials had been available in Colorado since 1966. The only difference from regular production Mustangs was the addition of the High Country Special decal.

The Mustang Sprint option consisted of GT stripes, pop-open gas cap and full wheel covers on six-cylinder Mustangs. V-8s got, in addition, the styled steel wheels with Wide Oval tires and the GT foglamps.

The rear taillight bezels on 1968 Mustangs are painted black, versus chrome on the 1967s.

The Reflective Group, optional on GT-equipped Mustangs, consisted of reflective GT side stripes and reflective paint on the wheels.

The Sports Trim Group consisted of woodgrain dash panel applique, knitted inserts in the bucket seats (hardtops and fastbacks) bright wheelwell moldings, two-tone louvered hood and, on V-8s only, argent styled steel wheels with E70x14 tires.

The primary difference between 1967 and 1968 Interior Decor Groups was the use of woodgrain dash appliques in 1968. The steering wheel center was redesigned for 1968, using a wide two-spoke center section.

1968 bucket seats have locking seat backs. A chrome lever is used to unlock the seat.

The collapsible spare tire was optional for the first time in 1968, as were front headrests.

Beginning with 1968, the disc brakes used on Mustangs have a single piston floating caliper.

1968 Fastback GT

Chapter 5

1969 Mustang

Production Figures

63A 2dr Fastback	56,022	65E 2dr Hardtop Grande	22,182
63B 2dr Fastback Deluxe	5,958	76A Convertible	11,307
63C 2dr Fastback Mach 1	72,458	76B Convertible Deluxe	3,439
65A 2dr Hardtop	118,613	Total	299,824
65B 2dr Hardtop Deluxe	5,210		
65C 2dr Hardtop, bench seats	4,131		
65D 2dr Hardtop Deluxe, bench seats	504		

Specials (included in above figures)

Boss 302	1,628
Boss 429	859
(includes two Boss Cougars)	

Serial Numbers
9F02Z100001
9 — Last digit of model year
F — Assembly plant (F-Dearborn, R-San Jose, T-Metuchen)
02 — Plate code for Mustang fastback (01-hardtop, 03-convertible)
Z — Engine code
100001 — Consecutive unit number

Location
Stamped on plate riveted on driver's side of instrument panel, visible through the windshield; stamped on left inner fender; warranty plate is riveted on rear face of driver's door.

Engine Codes
T — 200 ci 1V 6 cyl 115 hp
L — 250 ci 1V 6 cyl 155 hp
F — 302 ci 2V V-8 220 hp
G — 302 ci 4V V-8 (Boss) 290 hp
H — 351 ci 2V V-8 250 hp
M — 351 ci 4V V-8 290 hp
S — 390 ci 4V V-8 320 hp
Q — 428 ci 4V V-8 (CJ) 335 hp
R — 428 ci 4V V-8 (CJ-R) 335 hp
Z — 429 ci 4V V-8 (Boss) 375 hp

V-8 Distributors
302 ci 220 hp — C8AF-12127-E/manual, C8OF-12127-C/automatic
302 ci 290 hp — C9ZF-12127-F
351 ci 250 hp — C9OF-12127-M, N/manual, C9OF-12127-M, T/automatic

351 ci 290 hp — C9OF-12127-M, N/manual, C9OF-12127-M, T/automatic
390 ci 320 hp — C9AF-12127-K/manual, C7AF-12127-AC/automatic
428 ci 335 hp — C8OF-12127-H/manual, C8OF-12127-J/automatic
429 ci 375 hp — C9ZF-12127-U, D

V-8 Carburetors
302 ci 220 hp — C8AF-9510-BD/manual, C9AF-9510-A/automatic
302 ci 290 hp — C9ZF-9510-J(Holley R-4511)
351 ci 250 hp — C9ZF-9510-A/manual, C9ZF-9510-B/automatic
351 ci 290 hp — C9ZF-9510-C/manual, C9ZF-9510-D/automatic
390 ci 320 hp — C9ZF-9510-E/manual, C9ZF-9510-F/automatic
428 ci 335 hp — C9AF-9510-M(Holley R-4279)/manual, C9AF-9510-H(Holley R-4280)/automatic
429 ci 375 hp — C9AF-9510-S(Holley R-4456)

Steering Gear Ratios
SMB-D — 19.9:1
SMB-F — 16:1
SMB-K — 16:1

1969 Mustang Prices

	Retail
Hardtop, 65A	$2,618.00
Convertible, 76A	2,832.00
SportsRoof, 63A	2,618.00
Mach 1, 63C	3,122.00
Grande, 65E	2,849.00
250 ci 155 hp 6 cyl (NA Mach 1)	25.91
302 ci 220 hp V-8 (NA Mach 1)	105.00
Extra charge over 302 ci V-8 for:	
351 ci 250 hp V-8 (std. Mach 1)	58.34
351 ci 290 hp V-8 (except Mach 1)	84.25
Mach 1 over 351 ci 250 hp	25.91
390 ci 320 hp V-8 (except Mach 1)	158.08
Mach 1 over 351 ci 250 hp	99.74
428 ci 335 hp V-8 (except Mach 1)	287.53
Mach 1 over 351 ci 250 hp	224.12
428 ci 335 hp Ram Air Cobra Jet (CJ-R) V-8 (except Mach 1)	420.96
Mach 1 over 351 ci 250 hp	357.46
Boss 302 ci 4V 8 cyl engine	676.15
429 ci 4V Cobra Jet HO (Boss 429)	1208.35
SelectShift transmission, 6 cyl	191.13
302 & 351 ci V-8s	200.85
390 & 428 ci V-8s	222.08
4-speed manual transmission, 302 & 351 ci V-8s	204.64
4-speed manual transmission, 390 & 428 ci V-8s	253.92
Power front disc brake (NA 200 ci 6 cyl)	64.77
Power steering	94.95

Power convertible top	52.95
Convertible glass rear window	38.86
GT Equipment Group (NA Grande, 6 cyl or 302 V-8)	146.71
Tachometer (V-8 only)	54.45
Limited slip differential, 250 & 302 V-8	41.60
Traction-Lok differential (NA 6 cyl & 302 V-8)	63.51
Optional axle ratio	6.53
Intermittent windshield wipers	16.85
High-back bucket seats (NA Grande)	84.25
Color-keyed racing mirrors	19.48
Handling suspension (NA Grande or W200, 250 & 428)	30.64
Competition Suspension (428 only, std. Mach 1)	30.64
Power ventilation (NA w/SelectAire)	40.02
Electric clock (std. Mach 1, Grande)	15.59
Tinted windshield & windows	32.44
Speed control (V-8 & SelectShift)	73.83
Remote control outside mirror, lefthand	12.95
SelectAire air conditioner (NA 200 ci & 428 ci with 4-speed manual)	379.57
Push-button AM radio	61.40
AM/FM stereo radio	181.36
Stereosonic tape (AM radio required)	133.84
Rear seat speaker (hardtop & Grande)	12.95
Rear seat deck (SportsRoof & Mach 1)	97.21
Full-width front seat (hardtop, NA console)	32.44
Tilt-away steering wheel	66.14
Rim Blow deluxe steering wheel	35.70
Console	53.82
Interior Decor Group (NA Mach 1, Grande)	101.10
with color-keyed mirror option	88.15
Deluxe Interior Decor Group (SportsRoof & conv.)	133.44
with color-keyed mirror option	120.48
Deluxe seatbelts with warning light	15.59
Vinyl-covered roof (Grande & hardtop)	84.25
Wheel covers (NA Mach 1, GT, Grande, std. Exterior Decor Group)	21.38
Wire wheel covers (std. Grande, NA Mach 1, GT Group)	79.51
Wire wheel covers (Exterior Decor Group)	58.27
Exterior Decor Group (NA Mach 1, Grande)	32.44
Chrome styled steel wheels (std. Mach 1, NA Grande & 200 ci 6 cyl)	116.59
with GT Group	77.73
with Exterior Decor Group	95.31
Adjustable head restraints (NA Mach 1)	17.00
Visibility Group	11.16
Functional adjustable rear spoiler (Boss 302)	19.48
Sport Slats (Boss 302)	128.28
Trunk-mounted 85 amp battery (Boss 429)	32.44
Functional front air spoiler (Boss 429)	13.05
Shaker hood scoop (351 & 390 engines)	84.25

1969 Exterior Colors

Colors	Code
Raven Black	A
Royal Maroon	B
Black Jade	C
Acapulco Blue	D
Aztec Aqua	E
Gulfstream Aqua	F
Lime Gold	I
Wimbledon White	M
Winter Blue	P
Champagne Gold	S
Candyapple Red	T
Meadowlark Yellow	W
Indian Fire Red	Y
New Lime	2
Calypso Coral	3
Silver Jade	4
Pastel Grey	6

1969 Interior Trim

Trim	Code*
Black vinyl	2A
Blue vinyl	2B
Red vinyl	2D
Ivy Gold vinyl	2G
Nugget Gold vinyl	2Y
Black comfortweave, high buckets	4A(DA)
Red comfortweave, high buckets	4D(DD)
White comfortweave, high buckets	(DW)
Black comfortweave, luxury	5A
Blue comfortweave, luxury	5B
Red comfortweave, luxury	5D
Ivy Gold comfortweave, luxury	5G
White comfortweave, luxury	5W
Nugget Gold comfortweave, luxury	5Y
Black comfortweave, bench	8A(9A)
Blue comfortweave, bench	8B(9B)
Red comfortweave, bench	8D(9D)
Nugget Gold comfortweave, bench	8Y(9Y)
Black, convertible	7A
Blue convertible, deluxe	7B
Red convertible, deluxe	7D
Ivy Gold convertible, deluxe	7G
White convertible, deluxe	7W
Nugget Gold convertible, deluxe	7Y
Black cloth & vinyl, luxury	1A
Blue cloth & vinyl, luxury	1B
Ivy Gold cloth & vinyl, luxury	1G
Nugget Gold cloth & vinyl, luxury	1Y
Black, Mach 1	3A
Red, Mach 1	3D
White, Mach 1	3W

*Parentheses indicates Interior Decor Group

Convertible Top Colors
Black or White

1969 Mustang Facts

1969 was the second major restyle for the Mustang. Every dimension increased with the exception of wheelbase (it remained at 108 inches) and height, which was lowered by 1½ inches.

Including the two Boss engines, there was a total of ten different engines available.

Headlight configuration for the first time went to four, four-inch units. The fastback body style, called SportsRoof, came with simulated rear side scoops and a spoilered rear. Convertibles and hardtops used a simulated rear quarter panel vent.

The interior was totally restyled, with two separate dash pods available. The standard driver's side pod housed, from left to right, alternator, speedometer, a combination of fuel and temperature, and oil pressure. If the optional tachometer was ordered, it took the place of the fuel and temperature gauges. The temperature gauge was relocated to the far left, displacing the alternator gauge, while the temperature gauge displaced the oil pressure gauge on the far right.

The Deluxe Interior Group (standard on Mach 1, Grande and Boss 429) came with simulated woodgrain appliques on the dash, door panels and console. A clock was also housed on the passenger side dash pod. Numerals on the speedometer (and optional tachometer) were smaller than the standard interior's and the speedometer was divided in multiples of five. Instrument face color with the Deluxe Interior Group was dark grey, versus black for standard. This interior was optional only on SportsRoofs and hardtops.

The Interior Decor Group consisted of the molded door panels with woodgrain applique, comfortweave buckets, the Deluxe three-spoke Rim Blow steering wheel and a driver's side remote rectangular mirror. The high-back bucket seats were optional with either interior option.

1969 was the last year that bench seats were available and only on the hardtops. Also in 1969, the dual color-keyed race-type mirrors were optional on all Mustangs.

The Exterior Decor Group consisted of rocker panel moldings, wheelwell and rear end moldings.

The GT Equipment Group was still available on all three body styles as long as the engine was 351 ci or larger. It consisted of rocker panel GT stripes, GT gas cap, GT hubcaps on styled steel wheels, pin-type hood latches, heavy-duty suspension, simulated hood scoop and chrome quad outlets with dual-exhaust-equipped engines. Only 5,396 Mustangs got this option.

New engines included a 155 hp 250 ci six-cylinder, on which air conditioning was available. Two 351 ci V-8s, basically stretched 302s, joined the lineup. The two-barrel version was rated at 250 hp while the four-barrel was rated 290 hp. These engines were built at Ford's Windsor plant and thus are known as the 351 Windsor or 351W. The 390 was still available, but with a 470 cfm Ford carburetor rather than the previous Holley. A total of 10,494 Mustangs came with the 390.

The 428 Cobra Jet was the top production engine option. Two versions were available, both rated at 335 hp. The non-Ram Air version had the letter "Q" for its engine code, while those equipped with the functional Shaker hood scoop had the letter "R." The optional four-speed manual that was available with the 428CJ was the close ratio version, with a 2.32:1 first gear. All 428s also came with the larger 31 spline rear. 13,193 1969 Mustangs came with the 428CJ.

If a 3.91:1 (code V) or 4.30:1 (code W) rear axle ratio was ordered, the 428CJ was automatically upgraded to Super Cobra Jet (SCJ) status. These engines used special capscrew 427 LeMans type connecting rods, different crankshaft, flywheel and damper, and an external oil cooler mounted in front of the radiator, which reduced oil temperature by 30 degrees. All this for $6.53—without a doubt the best value option of the year.

Service bulletins on the 428CJ specified that an additional quart of oil be added during an oil change, for a total of six quarts.

The Grande Mustang was a luxury version of the Mustang hardtop. Standard equipment was the Deluxe Decor Group, wire wheel covers, color-keyed dual mirrors, two-tone paint stripes and Grande lettering on each C pillar.

The rare Mustang E was a specially equipped Mustang Sports-Roof designed for economy. It came with the 250 ci six-cylinder, high stall torque converter automatic transmission and a very low, 2.33:1 rear axle ratio. Mustang E lettering on the rear quarters identified the Mustang as such.

The Mach 1 took the place of the GT as the premier performance Mustang. It was based on the SportsRoof and came with a long list of standard features. The hood was painted flat black along with a similarly painted non-functional hood scoop. Reflective side and rear stripes were coordinated to complement the body color, as were color-keyed dual racing mirrors. Adding to the racer image were the NASCAR hood pin latches—a deleteable option. Chrome styled steel wheels and a chrome pop-open gas cap were also used. The Deluxe Decor Group was used in the interior. Standard engine was the 351-2V. The optional 4V engines came with chrome quad outlets. The Competition Suspension was standard equipment and included staggered rear shocks with the four-speed 428CJs.

Standard tires on the Mach 1 were E70x14s. F70x14 RWL Polyglas tires were mandatory with the 428s.

The Shaker scoop was optional on 351 2V, 351 4V and 390 engines. It was standard with the 428CJ-R.

The limited production Boss 302 was based on the SportsRoof body, but without the simulated side scoops. Flat black paint was used on the hood, headlight buckets, rear deck and taillight panels. A large C side stripe with Boss 302 lettering was used on the sides. Colors were limited to just four: Wimbledon White, Bright Yellow, Calypso Coral and Acapulco Blue. Most Boss 302s came with the standard black Mustang interior, though other colors were optional. A front chin spoiler was standard, while the rear window Sport Slats and rear wing were options. The rear spoiler was plastic on the 1969 Boss 302.

Standard wheels were argent painted 15x7 Magnum 500s using Goodyear F60x15 Polyglas tires. Chrome Magnum 500s were optional. All four fenders on the Boss 302 were radiused so that the tires would not hit.

The Boss 302 engine block was a special strengthened four-bolt main version of the production 302. It used forged steel connecting rods, a forged steel crank and special cylinder heads utilizing 2.23-inch (intake) and 1.72-inch (exhaust) valves. An aluminum high-rise intake manifold and a Holley 780 cfm carburetor provided induction. All Boss 302 engines came with a mechanical lifter camshaft and a dual-point distributor. All Boss 302s came with a four-speed wide ratio (2.78:1 first gear) manual transmission.

Other standard Boss 302 features were front disc brakes, quick ratio (16:1) steering, 3.50:1 rear axle ratio and staggered rear shocks.

The Boss 302's consecutive unit number was stamped on the engine block (at the rear center on a special pad). The Boss 302 and Boss 429 were the only production Mustangs where the original engine block can be matched to the original car.

The Boss 429 Mustang was a limited production Mustang designed to homologate the Boss 429 engine for NASCAR racing. All were built at the Kar Kraft facility in Brighton, Michigan. Partially completed SportsRoof Mustangs that originally were to receive the 428SCJ engine were modified to accept the large 429 engine. The suspension was lowered and moved further outwards one inch, using spindles and control arms unique to the Boss 429. Other features included Boss 429 fender decals, manually controlled hood scoop, a front spoiler that was shallower than the Boss 302 spoiler, color-keyed dual racing mirrors, engine oil cooler, trunk-mounted battery, power steering, power front disc brakes, close ratio four-speed manual transmission, 3.91:1 rear axle with Traction-Lok, ¾-inch rear sway bar, chrome 15x7 Magnum 500 wheels (long center cap) with F60x15 Goodyear RWL Polyglas GT tires, the Deluxe Decor interior, 8000 rpm tachometer and AM radio.

The Boss 429 engine was based on a strengthened version of the production 429. These blocks had HP429 cast into the front of the block (driver's side). It used four-bolt mains, a forged steel crank and forged steel connecting rods. The first 279 engines were tagged 820-S and came with NASCAR-type connecting rods (with ½-inch rod bolts) while all subsequent engines were tagged 820-T and used beefed up production rods. Cylinder heads were aluminum and featured a modified hemi-type combustion chamber which Ford called "crescent." These used the "dry-deck" method, meaning no head gaskets were used. Each cylinder, oil passage and water passage had its individual "O" ring to seal it. An aluminum intake manifold with a 735 cfm Holley carburetor provided induction. The camshaft was a hydraulic type. Early engines used magnesium valve covers while later ones were aluminum.

Each Boss 429 Mustang came with a KK sticker placed on the inside of the driver's door above the Ford warranty plate which signified Kar Kraft's production number. The first Boss 429 was numbered "KK NASCAR 1201" while the last 1969 was numbered

2059. Some Boss 429s may have this silver tape stripe missing; a small brass plate was substituted by Kar Kraft on a small number of cars.

The Boss 429's serial number was stamped on the back side of the engine block assembly, on the inner front fender panels, on the transmission housing and on the chassis itself.

1969 Boss 302

1969 SportsRoof GT

Chapter 6

1970 Mustang

Production Figures

63A 2dr Fastback	39,470	65E 2dr Hardtop	
65A 2dr Hardtop	77,161	Grande	13,581
76A Convertible	6,199	Total	190,727
63B 2dr Fastback	6,464	**Specials (included in above figures)**	
65B 2dr Hardtop	5,408		
76B Convertible	1,474	Boss 302	7,013
63C Fastback Mach 1	40,970	Boss 429	499

Serial Numbers
0F04F100001
0 — Last digit of model year
F — Assembly plant (F-Dearborn, R-San Jose, M-Metuchen)
04 — Plate code for Mustang (01-hardtop, 02-fastback, 03-convertible, 04-Grande, 05-Mach 1)
F — Engine code
100001 — Consecutive unit number

Location
Stamped on plate riveted to instrument panel on driver's side, visible through the windshield; a vehicle certification label, mounted on the rear face of the driver's door, replaced the previous warranty plate.

Engine Codes
T — 200 ci 1V 6 cyl 120 hp
L — 250 ci 1V 6 cyl 155 hp
F — 302 ci 2V V-8 220 hp
G — 302 ci 4V V-8 290 hp (Boss)
H — 351 ci 2V V-8 250 hp (351W & 351C)
M — 351 ci 4V V-8 300 hp
Q — 428 ci 4V V-8 335 hp (CJ)
R — 428 ci 4V V-8 335 hp (CJ-R)
Z — 429 ci 4V V-8 375 hp (Boss)

V-8 Distributors
302 ci 220 hp — D0AF-12127-T
302 ci 290 hp — C9ZF-12127-E
351 ci 250 hp — D0AF-12127-H/manual, D0AF-12127-AO/automatic
351 ci 300 hp — D0OF-12127-V/manual, D0OF-12127-Z/automatic
428 ci 335 hp — D0ZF-12127-C/manual, D0ZF-12127-G/automatic
429 ci 375 hp — C9ZF-12127-D

V-8 Carburetors
302 ci 220 hp — D0AE-9510-C/manual, D0AE-9510-D or V/automatic
302 ci 290 hp — D0ZF-9510-Z (Holley R-4653)
351 ci 250 hp — D0OF-9510-F/manual, D0OF-9510-L/automatic
351 ci 300 hp — D0OF-9510-AB/manual, D0OF-9510-AC/automatic
428 ci 335 hp — D0ZF-9510-AA or AD/manual(Holley R-4513, R-4515 [with A/C]), D0ZF-9510-AB or AC/automatic(Holley R-4514, R-4516 [with A/C])
429 ci 375 hp — D0OF-9510-S(Holley R-4647)

Steering Gear Ratios
SMB-D — 19.9:1
SMB-F — 16.1:1
SMB-K — 16.1:1

1970 Mustang Prices

	Retail
Hardtop, 65A	$2,721.00
SportsRoof, 63A	2,771.00
Convertible, 76A	3,025.00
Grande, 65E	2,926.00
Mach 1, 63C	3,271.00
Boss 302	3,720.00
250 CID 155 hp 6 cyl extra charge over 200 CID 6 cyl	39.00
302 CID 220 hp V-8 extra charge over 200 CID 6 cyl	101.00
351 CID 250 hp V-8 extra charge over 302 CID V-8	45.00
351 CID 300 hp V-8 extra charge over 302 CID V-8	93.00
Mach 1 over 351 CID 250 hp	48.00
428 CID 335 hp Cobra V8 extra charge over 302 CID V-8	356.00
Mach 1 over 351 CID 250 hp	311.00
428 CID 335 hp Cobra Jet Ram Air (except Mach 1)	421.00
Mach 1 over 351 CID 250 hp	376.00
Boss 429 CID 4V V-8	1,208.00
SelectShift Cruise-O-Matic (except 428)	201.00
with 428 V-8s	222.00
4-speed manual transmission, 302, 351, 428 V-8s	205.00
Power front disc brakes (NA 200 CID 6 cyl)	65.00
Power steering	95.00
SelectAire air conditioner (NA w/200 CID, Boss 302 428 CID w/4 speed manual)	380.00
Electric clock, rectangular (NA Grande, Mach 1 or Decor Group)	16.00
Electric clock, round, Decor Group only (std. Grande, Mach 1)	16.00
Console	54.00
Convenience Group, Grande, Mach 1, Boss 302 & Decor Group	32.00
All others	45.00
Rear window defogger, 2dr Hardtops	26.00
Color-keyed dual racing mirrors	26.00
Deluxe seatbelts with reminder light	15.00

Rear Sport Deck seat (SportsRoof)	97.00
Rim Blow Deluxe steering wheel	39.00
Tilt steering wheel	45.00
Intermittent windshield wipers	26.00
Space-saver spare (std. Boss 302, NA 200 CID 6 cyl)	
w/E70 or F70x14 WSW or RWL tires	7.00
w/E78x14 WSW tires	13.00
w/E78x14 BSW tires	20.00
AM radio	61.00
AM/FM radio	214.00
Stereosonic tape (AM radio required)	134.00
Front bumper guards	13.00
Decor Group, Boss 302 and other models	78.00
Convertibles	97.00
Rocker panel molding (std. Decor Group, NA Grande, Mach 1 or Boss 302)	16.00
Dual accent paint stripes, Mach 1	13.00
Vinyl roof (hardtop & Grande)	26.00
Sport Slats (requires dual racing mirrors)	65.00
Trim rings/hubcaps (std. Boss 302)	26.00
Wheel covers (std. Grande)	26.00
Sports wheel covers, Grande, Boss 302	32.00
All others	58.00
Wire wheel covers, Grande	53.00
All others	79.00
Magnum 500 chrome wheels, Boss 302 only	129.00
Argent style steel wheels, Grande (NC Mach 1)	32.00
All others	58.00
Drag Pack axle, 428 CID w/3.91 & 4.30 ratios	155.00
Optional ratio axle	13.00
Traction-Lok differential axle	43.00
Heavy-duty 55 amp battery	13.00
Heavy-duty 70 amp battery	13.00
Extra cooling package	13.00
Shaker hood scoop, Boss 302, 351	65.00
Rear deck spoiler, SportsRoof models only	20.00
Quick ratio steering (std. Boss 302)	16.00
Competition Suspension (NA 6 cyl)	31.00
Tachometer and trip odometer (NA 6 cyl)	54.00

1970 Exterior Colors

Colors	Code	Colors	Code
Raven Black	A	Medium Blue Metallic	Q
Dark Ivy Green Metallic	C	Medium Gold Metallic	S
		Red	T
Yellow	D	Grabber Orange	U
Medium Lime Metallic	G	Grabber Green	Z
Grabber Blue	J	Calypso Coral	1
Bright Gold Metallic	K	Light Ivy Yellow	2
Wimbledon White	M	Silver Blue Metallic	6
Pastel Blue	N		

1970 Interior Trim	Code	1970 Interior Trim	Code
Black vinyl	BA	Ginger Blazer Stripe cloth	CF
Blue vinyl	BB	Black Houndstooth cloth & vinyl	AA
Vermillion vinyl	BE	Blue Houndstooth cloth & vinyl	AB
Ginger vinyl	BF	Vermillion Houndstooth cloth & vinyl	AE
Ivy vinyl	BG	Ginger Houndstooth cloth & vinyl	AF
White vinyl	BW	Ivy Houndstooth cloth & vinyl	AG
Black comfortweave vinyl	EA	Black Mach 1 knitted vinyl	3A
Blue comfortweave vinyl	EB	Blue Mach 1 knitted vinyl	3B
Ivy comfortweave vinyl	EG	Ivy Mach 1 knitted vinyl	3G
White comfortweave vinyl	EW	White Mach 1 knitted vinyl	3W
Black comfortweave vinyl	TA	Vermillion Mach 1 knitted vinyl	3E
Blue comfortweave vinyl	TB	Ginger Mach 1 knitted vinyl	3F
Ivy comfortweave vinyl	TG		
White comfortweave vinyl	TW		
Vermillion Blazer Stripe cloth	UE		
Ginger Blazer Stripe cloth	UF		
Vermillion Blazer Stripe cloth	CE		

Convertible Top Colors
Black or White

1970 Mustang Facts

1970 Mustangs were mildly restyled. Headlight configuration reverted to a single seven-inch lamp on each side, this time inside the grille opening. Simulated scoops took the place of the outside headlights. The rear taillight panel was flat rather than concave and the taillights are recessed.

In the interior, a new steering wheel was used. The ignition switch was relocated on the steering column, as with all other domestic makes. The only optional interior was the Decor Group. It consisted of knitted vinyl or Blazer Stripe high-back buckets, simulated woodgrain appliques on the dash, deluxe steering wheel, molded door panels with simulated woodgrain appliques, dual color-keyed racing mirrors, and rocker panel and wheel lip moldings.

High-back bucket seats were standard equipment on all Mustangs. The Sport Slats and rear deck wing were optional on all Mustang SportsRoofs.

The Shaker scoop was again optional on engines other than the 428 CJ-R. It was available with both 351s and also on the Boss 302.

The Competition Suspension was unchanged, except for the addition of a rear stabilizer bar, ½ inch on 351s and ⅝ inch on 428 engines.

The Drag Pack option, available only with 428 engines, came with 3.91:1 or 4.30:1 axle ratios. It consisted of an engine oil cooler, stronger 427-type connecting rods and different harmonic balancer/flywheel combinations. This option changes a Cobra Jet to a Super Cobra Jet.

All 1970 Mustangs with dual exhaust systems benefited from a new exhaust system that used two mufflers mounted ahead of the rear axle, replacing the transverse muffler of previous years. All four-speed equipped Mustangs used a Hurst shifter.

While the 390 was dropped, a new 351 joined the line-up. The 351 Cleveland or 351C, available in two-barrel and four-barrel forms (250 and 300 hp) used cylinder heads which were very similar to those used on the Boss 302. The two-barrel engines used the so-called two-barrel heads which have smaller ports and valves. The four-barrel heads have the same size valves and ports as the Boss 302 cylinder heads. The 351W 4V was dropped but Mustangs equipped with the two-barrel 351 could either be a 351W or a 351C.

The Mach 1 was restyled, getting a revised grille that had two driving lamps. The NASCAR hood pins were replaced with twist-type latches and the blacked-out hood treatment of 1969 was replaced by a stripe arrangement in the middle of the hood, painted white or black. A complementary stripe was used on the rear deck lid, while the rear taillight panel got a honeycomb treatment. On the sides, aluminum rocker panel moldings with large Mach 1 lettering replaced the previous year's side tape stripes. Oval exhaust extensions replaced the quad arrangement on Mach 1s (and other Mustangs) with dual exhausts. Simulated sports wheel covers replaced the chrome styled steel wheels, but painted styled steel wheels were a no-cost option.

The Grande, too, was relatively unchanged. A half vinyl roof became part of the Grande package while in the interior, hound's-tooth cloth was used on the seats. The wire wheel covers became optional on the 1970 Grande.

The Boss 302 got a new tape stripe treatment that began on the hood (similar to the Mach 1), but then continued along the rear of the hood, down the front fenders and across the sides to the rear. Standard wheel size remained at 15x7; however, a trim ring/hubcap arrangement took the place of the Magnum 500s. Chrome Magnum 500 wheels, though, were optional but only on the Boss 302. The Shaker scoop was optional on the Boss 302.

Mechanically, the Boss 302 benefited from the redesigned dual exhaust system, the standard Hurst shifter, and the Competition Suspension which used a rear stabilizer bar measuring ⅝ inch. The engine got smaller intake valves, 2.19 inches versus 2.23 inches for 1969, which resulted in slightly better response. Aluminum valve covers replaced the chrome steel ones of 1969.

The 1970 Boss 429 featured a gloss black hood scoop, chrome Magnum 500 wheels that used a Boss 302 type center cap, the revised exhaust system and the Hurst shifter. Color availability was

Grabber Blue, Grabber Orange, Grabber Green, Calypso Coral and Pastel Blue. Interior color selection was limited to black or white.

The 1970 Boss 429 engine was the 820-T version; however, it was fitted with a mechanical lifter camshaft and a more efficient radiator fan. Some engines were tagged 820-A. These were 820-T engines that had some minor emission system modifications. The ¾ inch rear stabilizer bar was replaced by the ⅝ inch unit found on other Competition Suspension-equipped Mustangs.

A total of 499 1970 Boss 429s were built. KK numbers range from KK2060 to KK2558.

As with 1969 Bosses, all 1970 Boss 302 Mustangs had the consecutive unit number on the engine block for additional identification. The Boss 429s had additional VIN numbers on the engine block, transmission and inner fenders.

1970 Boss 429

Chapter 7

1971 Mustang

Production Figures
65D 2dr Hardtop	65,696
63D 2dr SportsRoof	23,956
76D Convertible	6,121
65F 2dr Hardtop Grande	17,406
63F 2dr SportsRoof Mach 1	36,499
Total	149,678

Specials (included in above figures)
Boss 351	1,806

Serial Numbers
1F01M100001
1 — Last digit of model year
F — Assembly plant (F-Dearborn, T-Metuchen)
01 — Plate code for Mustang body (01-2dr hardtop, 02-2dr SportsRoof, 03-convertible, 04-2dr hardtop Grande, 05-2dr SportsRoof Mach 1)
M — Engine code
100001 — Consecutive unit number

Location
Stamped on plate attached to driver's side of instrument panel, visible through the windshield; certification label is attached to rear face of driver's door.

Engine Codes
L — 250 ci 1V 6 cyl 145 hp
F — 302 ci 2V V-8 210 hp
H — 351 ci 2V V-8 240 hp
M — 351 ci 4V V-8 280 hp (CJ)

Engine Codes
M — 351 ci 4V V-8 285 hp
R — 351 ci 4V V-8 330 hp (Boss)
C — 429 ci 4V V-8 370 hp (CJ)
J — 429 ci 4V V-8 375 hp (CJ-R)

V-8 Distributors
302 ci 210 hp — D0AF-12127-Y/manual,
  D0OF-12127-AC/automatic
351 ci 240 hp — D0OF-12127-T/manual,
  D0OF-12127-U/automatic
351 ci 285 hp — D0OF-12127-V/manual,
  D0OF-12127-G/automatic
351 ci 330 hp — D0OF-12127-V(70F118)D-12
429 ci 370 hp — D0OF-12127-AA/manual,
  D0OF-12127-NA/automatic
429 ci 375 hp — D0OF-12127-AA(70F56)D-12/manual
  D0OF-12127-NA(71F117)D-12/automatic

V-8 Carburetors
302 ci 210 hp — D1AF-9510-BA, DA or D1OF-9510-ABA/
  manual, D1AF-9510-DA, AA, SA, TA or D1OF-9510-ABA/
  automatic

351 ci 240 hp — D1OF-9510-PA, RA, YA, ZA, D, MF, KA or D1ZF-9510-SA, UA/manual D1MF-9510-KA, D1OF-9510-RA, YA, ZA or D1ZF-9510-SA, UA/automatic

351 ci 280 hp — D1ZF-9510-ZA/all

351 ci 285 hp — D1OF-9510-EA/manual, D1OF-9510-AAA, FA/automatic

351 ci 330 hp — D1ZF-9510-FA or D0ZF-9510-Z

429 ci 370 hp — D0OF-9510-A/manual, D0OF-9510-B/automatic

429 ci 375 hp — D1OF-9510-SA or D1ZF-9510-YA(Holley R-6127)/manual, D1OF-9510-TA or D1ZZ-9510-XA(Holley R-6128)/automatic

1971 Mustang Prices

	Retail
2dr Hardtop	$2,911.00
2dr SportsRoof	2,973.00
Convertible	3,227.00
2dr Hardtop Grande	3,117.00
2dr SportsRoof Mach 1	3,268.00
2dr SportsRoof Boss 351	4,124.00
302 2V 210 hp	N/C
351 2V 240 hp	45.00
351 4V 285 hp	93.00
429 4V 370 hp*	372.00
429 4V 370 hp Ram Air*	436.00
Optional axle ratio (NA Boss 351)	13.00
Heavy-duty 70 ampere battery	16.00
Traction-Lok differential (std. Boss 351)	45.00
Drag Pack, 429s only 3.91 & 4.30 w/Traction-Lok	155.00
4.30 w/No-Spin Detroit Locker	207.00
Extra cooling package	14.00
Instrumentation Group Mach 1 w/console	37.00
Mach 1 w/o console	54.00
Grande w/o console	62.00
All others	79.00
Dual Ram Air Induction (NA 250 & 302 engines)	65.00
Competition Suspension	31.00
Rear deck spoiler (SportsRoof only)	32.00
4-speed manual, V-8 engines	216.00
SelectShift Automatic, 250, 302 & 351 engines	217.00
with 429 engine	238.00
Power front disc brakes	70.00
Power steering	115.00
Power windows	127.00
AM radio	66.00
AM/FM stereo radio	214.00
AM/stereo tape system (requires AM radio)	134.00
SelectAire air conditioner	412.00
Front & rear bumper guards (NA Mach 1)	31.00
Console (includes clock), Grande & Mach 1	60.00
All others	76.00
Protection Group, models w/bumper guards	34.00

All others	45.00
Sport Deck seat	97.00
Rim Blow steering wheel	39.00
Tilt-away steering wheel (requires power steering)	45.00
Trim, Decor Group, Convertible & Boss 351	97.00
All others	78.00
Sports Interior, Mach 1 & SportsRoof	130.00
Boss 351	88.00
Bodyside tape stripe (Mach 1 only, std. Boss 351)	26.00
Vinyl roof, 2dr Hardtop (std. Grande)	89.00
Base wheel covers	26.00
Sport wheel covers, Grande	32.00
Mach 1	23.00
All others	58.00
Trim rings/hubcaps, Grande	9.00
All others	35.00
Magnum 500 chrome wheels (requires Competition Suspension, F60x15 RWL tires, & includes Space-saver spare), Grande	129.00
Boss 351 & Mach 1	120.00
All others	155.00
Intermittent windshield wipers	26.00
Rear window electric defroster	48.00
Color-keyed dual racing mirrors	26.00
Deluxe seatbelts	17.00
Complete tinted glass, Convertible	15.00
All others	40.00
Convenience Group	51.00
Tires	
E78x14 WSW	32.00
E70x14 Wide Oval WSW over E78x14	39.00
F70x14 Wide Oval WSW over E78x14	68.00
F60x15 Wide Oval BSW/WL over E78x14	99.00
F70x14 Wide Oval BSW/WL over E70x14 WSW	81.00
F70x14 Wide Oval WSW over E70x14 WSW	29.00
F70x14 Wide Oval BSW/WL over E70x14 WSW	42.00
F60x15 Wide Oval BSW/WL over E70x14 WSW	60.00
F70x14 Wide Oval BSW/WL over F70x14 WSW	13.00
F60x15 Wide Oval BSW/WL over F70x14 WSW	31.00
F60x15 Wide Oval BSW/WL over F70x14 BSW/WL	18.00

*Includes 80 amp heavy-duty battery, Competition Suspension, 55 amp alternator, extra cooling package, dual exhausts, bright engine dress-up with cast aluminum rocker covers, Mach 1 hood, plus a 3.50 non-locking axle with Ram Air and a 3.25 non-locking axle with non-Ram Air. Requires optional transmission, & Drag Pack when 3.91 or 4.30 axle is ordered plus Ram Air: F70x14 Wide Oval belted WSW tires, F70x14 BSW/WL on Mach 1. non-Ram Air: E70x14 Wide Oval belted WSW tires, except Mach 1 Air conditioner not available with Drag Pack and Ram Air.

1971 Exterior Colors

Color	Code
Raven Black	A
Maroon Metallic	B
Dark Ivy Green Metallic	C
Grabber Yellow	D
Medium Yellow Gold	E
Grabber Lime	I
Grabber Blue	J
Wimbledon White	M
Pastel Blue	N
Medium Green Metallic	P
Light Pewter Metallic	V
Grabber Green Metallic	Z
Bright Red	3
Medium Brown Metallic	5
Silver Blue Metallic	6
Light Gold	8
Gold Metallic	Special
Gold Glamour	Special

1971 Interior Trim

Trim	Code
Black vinyl	1A
Medium Blue vinyl	1B
Vermillion vinyl	1E
Medium Ginger vinyl	1F
Medium Green vinyl	1R
White vinyl	1W
Black knitted vinyl	3A
White knitted vinyl	3W
Vermillion cloth & vinyl	2E
Medium Ginger cloth & vinyl	2F
Medium Blue cloth & vinyl	2B
Medium Green cloth & vinyl	2R
Black knitted vinyl	5A
White knitted vinyl	5W
Vermillion knitted vinyl	5E
Medium Blue knitted vinyl	5B
Medium Green knitted vinyl	5R
Medium Ginger knitted vinyl	5F
Black knitted vinyl	CA
White knitted vinyl	CW
Vermillion knitted vinyl	CE
Medium Blue knitted vinyl	CB
Medium Green knitted vinyl	CR
Medium Ginger knitted vinyl	CF
Black cloth & vinyl	4A
Medium Blue cloth & vinyl	4B
Vermillion cloth & vinyl	4E
Medium Ginger cloth & vinyl	4F
Medium Green cloth & vinyl	4R

1971 Mustang Facts

While maintaining a basic resemblance to previous Mustangs, the 1971 Mustangs were the largest and heaviest yet. Every dimension increased; Wheelbase was stretched to 109 inches.

Engine choice still remained high; a total of ten different engines were available on the Mustang.

Base engine was the 250 ci six-cylinder on all models except the Mach 1 which came with the 210 hp 302 and the Boss 351 which got a 330 hp version of the 351.

Optional V-8s were the 210 hp 302, 240 hp 351, 285 hp 351, 370 hp 429 and 375 hp 429. All 351 V-8s were 351Cs. In May 1971, a low-compression 351CJ replaced the 285 hp 351 that was available at the beginning of the model year, rated at 280 hp. Both of these 351s had the same engine code, M.

The 428CJ was replaced by the 429CJ as the top Mustang engine option. The 429 belonged to the 385 Engine Series and as such, no parts were interchangeable with the 428. Wider, larger and heavier, the 429 would not readily fit into the 1970 Mustang engine compartment, which is one of the reasons that the 1971 Mustang got bigger. The bottom end and cylinder block were a variation of the 429/460 block, on which the Boss 429 was also based. The cylinder heads were similar to the 351 Cleveland in design. You could describe the 429 as a large Boss 302. As equipped in the Mustang, the 429CJ came with four-bolt mains, forged rods and pistons, 11.3:1 compression ratio, a hydraulic cam (similar to the Boss 429's), very large ports and valves, 2.25 inch intake and 1.72 inch exhausts. Regular production 429/460 engines came with cylinder heads that had smaller ports and valves. The 429CJ came with a Rochester Quadrajet four-barrel carburetor, and some early units had adjustable valve trains. All 429CJs came with aluminum valve covers.

The 429CJ became a 429SCJ if the Drag Pack option was ordered. It consisted of a 3.91:1 or 4.30:1 rear axle ratio with Traction-Lok or a 4.11:1 ratio with a Detroit Locker rear. Both of these engines had the engine code C. Some Mustangs with the Drag Pack also included an external engine oil cooler, though of a different design than previously used. The SCJ engine came with a 780 cfm carburetor rather than the Rochester Quadrajet, a mechanical lifter camshaft and adjustable rocker arms.

Mustangs equipped with the 429CJ Ram Air engine got the letter J for the engine code in the VIN. It was rated at 375 hp. The 429CJ-R engine could either be a CJ or SCJ, if it had the Drag Pack option or not.

All four-speed manual transmissions came with a Hurst shifter.

The Competition Suspension, available only on Mustangs with 351-2V engines and larger, consisted of heavier-duty shock absorbers, springs and front and rear stabilizer bars (rear bars for 351 four-barrel and larger engines only), and staggered rear shocks (except for the 351-2V engine). Mustangs with this option got power steering with variable ratio. The Competition Suspension was also mandatory with the optional 15x7 chrome Magnum 500 wheels.

The Dual Ram Induction option consisted of a hood with two functional NASA-type hood scoops, which were controlled by engine vacuum. Twist-type hood locks, a Tu-tone paint treatment, black or argent, and Ram Air decals rounded out the option. It was available only on 351 and larger engines.

The optional variable ratio steering had a 15.7:1 ratio.

While the rear deck spoiler remained on the option list, the Sport Slats were deleted, due to the low angle of the SportsRoof's rear window.

For the first time, power windows and an electric rear window defroster were optional. The electric defroster was not available on convertibles.

The four-pod instrument panel of 1969-70 was replaced by a three-pod design. Two large pods with a smaller one in between them dominated the driver's side of the dash. The large left pod consisted of four warning lights: oil pressure, temperature, brakes and alternator. The center pod was for fuel while the large right unit housed the speedometer.

The optional instrumentation group (standard on the Boss 351 and not available with the 250 ci six-cylinder) consisted of three instruments, oil pressure, alternator and temperature, housed in a panel located above the radio. The warning lights on the left large pod were displaced by an 8000 rpm tachometer.

The three-spoke Rim Blow steering wheel was optional. Similar to the one used on 1969-70 Mustangs, it used a redesigned center pad. The tilt steering is optional; however, power steering was a mandatory option.

The Decor Group (not available on the Mach 1 or Grande) consisted of either knitted vinyl or cloth inserts for the seats, black instrument panel faces, the deluxe two-spoke steering wheel, molded door panels (available only on the convertible and Boss 351), rear ashtray, dual color-keyed racing mirrors, and rocker panel and wheel lip moldings. The moldings were not available with the Boss 351.

The Mach 1 Sports Interior, available on the Mach 1 and all other SportsRoofs, consisted of knitted vinyl seats, the two-spoke deluxe steering wheel, molded door panels, black dash panel applique, woodgrain applique for the center instrument panel, rear ashtray, electric clock, the instrumentation group, bright pedal pads, and color-keyed rubber floor mats stitched directly on the carpets (front only).

The fold-down rear seat option included a Space-saver spare tire.

The Mach 1 for 1971 used a different grille-bumper combination. The honeycomb grille housed two driving lamps while the front bumper was covered with urethane color-keyed to the Mustang's paint. Fender moldings were also color-keyed. The NASA hood was standard equipment (non-functional). The hood along with the lower body were painted either black or argent, depending on body paint. Color-keyed dual racing mirrors were standard and all Mach 1s came with front fender Mach 1 decals. There was also a small Mach 1 decal on the rear deck above the pop-open gas cap. Boss 351 side stripes were optional.

Standard engine with the Mach 1 was the 302 V-8, and all other V-8s were optional, except the 330 hp 351 found on the Boss 351.

The Boss 351 replaced the Boss 302 and the Boss 429 as the premier performance Mustang. Like previous Boss Mustangs, it was a complete package, with limited options. Standard was the 330 hp 351 Cleveland featuring a four-bolt main block, large port cylinder heads and valves, a solid lifter camshaft, an 11.7:1 compression ratio and aluminum valve covers. Other standard features were Ram Air, 3.91:1 rear axle with Traction-Lok, four-speed manual transmission, Competition Suspension, power front disc brakes, front spoiler, the Mach 1 front grille and lower body side paint treatment, bodyside tape treatment and Boss 351 decals in place of the Mach 1 decals. The Boss 351, however, came with the standard chrome front bumper. The hood differed from the Mach 1's as the black or argent paint covered most of the hood.

Standard wheels on the Boss 351 were 15x7 with trim rings/hubcaps. Optional were the 15x7 chrome Magnum 500s. Tires in both cases were Goodyear F60x15 RWL.

The Grande was still promoted as the luxury Mustang, available only on the hardtop body style. A full vinyl roof, Grande lettering on the "C" pillar, dual accent paint stripe, color-keyed racing mirrors, bright rocker panel and wheel lip moldings, and special Grande wheel covers were all standard equipment on the exterior. In the interior, the Grande came with the Deluxe two-spoke steering wheel, black dash panel appliques, woodgrain appliques in the center dash, molded door panels, Lambeth cloth seat inserts, electric clock, rear ashtray and bright trim on the pedals. All Mustang engines were available on the Grande.

Further watering down the overall performance image of the Mach 1 and Boss 351 Mustangs was the availability of the Sports Hardtop option late in the model year. Based on the hardtop body, it used the Mach 1's honeycomb grille and color-keyed bumper, the standard Mach 1 hubcap/trim rings, the non-functional NASA hood and the Boss 351's side stripes.

1971 Mach 1

1971 Boss 351

Chapter 8

1972 Mustang

Production Figures

65D 2dr Hardtop	57,350	63R 2dr SportsRoof	
63D 2dr SportsRoof	15,622	Mach 1	27,675
76C Convertible	6,121	Total	125,093
65F 2dr Hardtop Grande	18,045		

Serial Numbers

2F05Q100001

2 — Last digit of model year
F — Assembly plant (F-Dearborn)
05 — Plate code for Mustang (01-2dr hardtop, 02-2dr SportsRoof, 03-convertible, 04-2dr hardtop Grande, 05-2dr SportsRoof Mach 1)
Q — Engine code
100001 — Consecutive unit number

Location

Stamped on plate riveted to driver's side of dash, visible through the windshield; certification label attached to rear face of driver's door.

Engine Codes

L — 250 ci 1V 6 cyl 98 hp
F — 302 ci 2V V-8 140 hp
H — 351 ci 2V V-8 177 hp
Q — 351 ci 4V V-8 266 hp (CJ)
R — 351 ci 4V V-8 275 hp (HO)

1972 Mustang Prices

	Retail
6 cyl Models	
2dr Hardtop, 65D	$2,679.00
2dr SportsRoof, 63D	2,736.00
2dr Convertible, 76D	2,965.00
Grande Hardtop, 65F	2,865.00
8 cyl Models	
2dr Hardtop, 65D	2,766.00
2dr SportsRoof, 63D	2,823.00
2dr Convertible, 76D	3,051.00
Grande Hardtop, 65F	2,952.00
Mach 1 SportsRoof, 63R	3,003.00
351 cid 2V 8 cyl	40.79
351 cid 4V 8 cyl (includes NASA hood)	115.44
351 cid 4V HO 8 cyl, with Mach 1	783.00
All others	812.00
SelectShift Cruise-O-Matic transmission	203.73
4-speed manual with Hurst shifter	192.99
SelectAire air conditioner	367.59

Axle, optional ratio	11.66
Axle, Traction-Lok differential	42.64
Battery, heavy-duty 70 ampere	13.52
Belts, deluxe	15.49
Bumper guards, front & rear	28.19
Console, Grande & Mach 1 Sports Interior	53.40
all others	67.95
Convenience group	45.53
Decor group	69.79
Door edge guards	5.78
Electric defroster, rear window	42.64
Emission system, Calif.	13.87
Extra cooling package	12.59
Complete tinted glass, Convertible	13.52
All other models	35.94
Hood, NASA-type (std. Mach 1 with 351)	N/C
Instrumentation Group, Grande without console	55.24
All others	70.83
Mirrors, outside color-keyed dual racing	23.23
Paint, color-glow	34.90
Power front disc brakes	62.05
Power side windows	113.48
Power steering	102.85
Protection package	52.06
Radio, AM	59.17
Radio, AM/FM stereo	191.01
Ram Induction NASA hood (with 351 2V only)	58.24
Roof, vinyl	79.51
Roof, ¾ vinyl	52.35
Seat, Sport Deck rear	86.32
Spoiler, rear deck (SportsRoof models only)	29.12
Sports Interior option, Mach 1	115.44
Steering wheel, Rim Blow, Deluxe three-spoke	34.90
Steering wheel, tilt	40.79
Stereosonic tape system	120.29
Suspension, competition	28.19
Tape stripe, black or argent bodyside	23.23
Trim rings/hubcaps, Grande	7.86
All others	31.08
Wheel covers	23.23
Wheel covers, Sports, Grande	56.18
Mach 1 & Decor Group	48.33
All others	79.40
Wheels, Magnum 500 Chrome, Grande	115.44
Mach 1 & Decor Group	107.59
All others	138.67
Windshield wipers, intermittent	23.23
Glass, tinted windshield	22.42
Models having (5) E78x14 BSW tires, extra charge for:	
(5) E70x14 WSW	35.28
(5) F70x14 WSW	61.78
(5) F70x14 B/WL	73.57

(4) F60x15 B/WL (includes F78x14 Space-saver spare)	90.96
Models having (5) E70x14 WSW tires, extra charge for:	
(5) F70x14 WSW	26.51
(5) F70x14 B/WL	38.36
(4) F60x15 B/WL (includes F78x14 Space-saver spare)	55.68

1972 Exterior Colors

Color	Code
Wimbledon White	9A
Bright Red	2B
Medium Yellow Gold	6C
Bright Lime	4E
Grabber Blue	3F
Medium Brown Metallic	5H
Bright Blue Metallic	3J
Medium Green Metallic	4P
Dark Green Metallic	4Q
Maroon	2J
Light Blue	3B
Medium Lime Metallic	4F
Light Pewter Metallic	5A
Medium Bright Yellow	6E
Gold Glow	6F
Ivy Glow	4C

1972 Interior Trim

Trim	Code
Medium Ginger cloth & vinyl	4F
Vermillion cloth & vinyl	4E
Medium Blue cloth & vinyl	4B
Medium Green cloth & vinyl	4R
Black cloth & vinyl	4A
Medium Blue knitted vinyl	5B
Black knitted vinyl	5A
Vermillion knitted vinyl	5E
White knitted vinyl	5W
Medium Green knitted vinyl	5R
Medium Ginger knitted vinyl	5F
Vermillion vinyl	1E
White vinyl	1W
Black vinyl	1A
Medium Blue vinyl	1B
Medium Green vinyl	1R
Medium Ginger vinyl	1F
Vermillion cloth & vinyl	2E
Medium Ginger cloth & vinyl	2F
Medium Blue cloth & vinyl	2B
Medium Green cloth & vinyl	2R
Black knitted vinyl	CA
White knitted vinyl	CW
Vermillion knitted vinyl	CE
Medium Blue knitted vinyl	CB
Medium Green knitted vinyl	CR
Medium Ginger knitted vinyl	CF

Convertible Top Colors

Black or White

1972 Mustang Facts

The 1972 Mustangs were externally similar to the '71s. You could tell them apart by the Mustang script lettering on the deck lid above the right taillights—1971 Mustangs had Mustang lettering that covered the width of the deck lid. All Mustangs came with chrome rocker panel and wheel lip moldings except for the Mach 1. These were previously optional.

The Exterior Decor Group, available only on the standard hardtops and convertibles, gave these Mustang models more of a performance look. It consisted of the Mach 1 honeycomb grille and sportslamps, color-keyed front bumper, hood and fender moldings, lower bodyside paint treatment and the trim ring/hubcap arrangement for the wheels. The Exterior Decor Group could be combined with the Mach 1 tape bodyside stripes to give Mustangs so equipped the look of performance.

The base engine was the 250 ci six-cylinder. The standard Mach 1 engine was still the 302 ci V-8. Optional were two 351s, a 2V version rated at 177 hp and the 4V 351CJ rated at 266 hp.

The 351 HO was available for a brief time on all 1972 Mustang body styles. As such, it represented Ford's last true 1960s-style muscle engine/package. The 351 HO was essentially a low compression Boss 351 engine, pumping out 275 hp. About 1,000 or so of these engines were installed in Mustangs and they all came with a four-speed manual transmission, 3.91 Traction-Lok rear axle, dual exhausts, Competition Suspension, power front disc brakes and an electronic rev limiter. Tires were F60x15 on the chrome Magnum 500 wheels. No special identification was used other than a 351 HO decal on the air cleaner lid.

Two special Mustang Sprint packages were made available on Mustang SportsRoof and hardtop models in February 1972, to coincide with similar Pintos and Mavericks. Sprint Package A consisted of the Exterior Decor Group, dual color-keyed racing mirrors, trim rings/hubcap combination with E70x14 WSW tires. What made the Sprints stand out were their white paint with dual blue and red hood stripes, blue and red lower body treatment and a blue and red rear taillight panel. A USA shield was used on the rear fenders while the interior, too, was unique with white vinyl interior. Seats were white vinyl with blue Lambeth cloth.

Sprint Package B was identical but added the Competition Suspension and the chrome Magnum 500 wheels/F60x15 tire combination.

Fifty Sprint Package A convertibles were built for the Washington D.C. area for participation in the Cherry Day Parade.

1972 SportsRoof Sprint

Chapter 9

1973 Mustang

Production Figures

63D 2dr SportsRoof	10,820	63R 2dr SportsRoof	
65D 2dr Hardtop	51,480	Mach 1	35,440
76D Convertible	11,853	Total	134,867
65F 2dr Hardtop Grande	25,274		

Serial Numbers

3F03H100001

3 — Last digit of model year
F — Assembly plant (F-Dearborn)
03 — Plate code for Mustang (01-2dr hardtop, 02-2dr SportsRoof, 03-convertible, 04-2dr hardtop Grande, 05-2dr SportsRoof Mach 1)
H — Engine code
100001 — Consecutive unit number

Location

Stamped on plate riveted to driver's side of dash, visible through the windshield; certification label attached to rear face of driver's door.

Engine Codes

L — 250 ci 1V 6 cyl 99 hp
F — 302 ci 2V V-8 141 hp
H — 351 ci 2V V-8 177 hp
Q — 351 ci 4V V-8 266 hp (CJ)

1973 Mustang Prices   Retail

6 cyl Models	
2dr Hardtop, 65D	$2,760.00
2dr SportsRoof, 63D	2,820.00
2dr Convertible, 76D	3,102.00
2dr Hardtop, Grande, 65F	2,946.00
8 cyl Models	
2dr Hardtop, 65D	2,847.00
2dr SportsRoof, 63D	2,907.00
2dr Convertible, 76D	3,189.00
2dr Hardtop, Grande, 65F	3,033.00
2dr SportsRoof Mach 1, 63R	3,088.00
351 cid 2V 8 cyl	40.79
351 cid 4V 8 cyl	107.00
SelectShift Cruise-O-Matic	203.73
4-speed manual with Hurst shifter	192.99
Air conditioner, SelectAire	367.59
Axle, optional ratio	11.66

Axle, Traction-Lok differential	42.64
Battery, heavy-duty 70 ampere	13.52
Belts, Deluxe	15.49
Bumper Group, Deluxe	25.00
Bumper Guards, rear	14.00
Console, Grande	53.40
All others	67.95
Convenience Group	45.53
Decor Group	51.00
Door edge guards	5.78
Electric rear window defroster	57.00
Emissions testing, Calif.	13.87
Extra cooling package	12.59
Floor mats, front color-keyed	13.30
Complete tinted glass, Convertible	13.52
All others	35.94
Hood, NASA-type	N/C
Instrumentation Group, Grande without console	55.24
All others	70.83
Mirrors, outside color-keyed dual racing	23.23
Paint, Metallic Glow	34.90
Paint, Tu-tone hood (NA w/Dual Ram Induction)	
Mach 1	18.00
All others	34.00
Power front disc brakes	62.05
Power side windows	113.48
Power steering	102.85
Protection Group, Grande	23.38
All others	36.00
Radio, AM	59.17
Radio, AM/FM stereo	191.01
Dual Ram Induction option (w/351 cid 2V only)	58.24
Roof, vinyl	79.51
Roof, ¾ vinyl	52.35
Seat, Sport Deck rear	86.32
Spoiler, rear deck (SportsRoof & Mach 1)	29.12
Sports Interior option, Mach 1	115.44
Steering wheel, leather-wrapped	23.10
Steering wheel, Rim Blow, Deluxe 3-spoke	34.90
Steering wheel, tilt	40.79
Stereosonic tape system	120.29
Suspension, competition	28.19
Tape stripe, black or argent bodyside	23.23
Trim rings/hubcaps, Grande	7.86
All others	31.08
Wheel covers	23.23
Wheel covers, sports, Grande	56.18
Mach 1 & Decor Group	48.33
All others	79.40
Wheels, forged aluminum, Grande	118.77
Mach 1 & Decor Group	110.92
All others	142.00

Windshield wipers, interval	23.23
Glass, tinted windshield	22.42
Models having (5) E78x14 BSW tires, extra charge for:	
(5) F78x14 BSW	17.00
(5) E70x14 WSW	35.28
(5) F70x14 WSW	61.78
(5) F70x14 B/WL	73.57
(5) GR78x14 steel-belted radial ply BSW	115.00
(5) GR78x14 steel-belted radial ply WSW	144.00
Models having (5) F78x14 tires, extra charge for:	
(5) F70x14 WSW	46.00
(5) F70x14 B/WL	57.00
(5) GR78x14 steel-belted radial ply BSW	98.00
(5) GR78x14 steel-belted radial ply WSW	127.00
Models having E70x14 WSW tires, extra charge for:	
(5) F70x14 WSW	26.51
(5) F70x14 B/WL	38.36
(5) GR78x14 steel-belted radial ply BSW	69.00
(5) GR78x14 steel-belted radial ply WSW	98.00

1973 Exterior Colors

Color	Code
Wimbledon White	9A
Bright Red	2B
Medium Yellow Gold	6C
Medium Blue Metallic	3D
Medium Brown Metallic	5H
Blue Glow	3K
Medium Copper Metallic	5M
Medium Aqua	4N
Medium Green Metallic	4P
Dark Green Metallic	4Q
Saddle Bronze Metallic	5T
Light Blue	3B
Medium Bright Yellow	6E
Ivy Glow	4C
Bright Green Gold Metallic	4B
Gold Glow	6F

1973 Interior Trim

Trim	Code
Black vinyl	AA
Medium Blue vinyl	AB
Medium Ginger vinyl	AF
Avocado vinyl	AG
White vinyl	AW
Black knitted vinyl	CA
Medium Blue knitted vinyl	CB
Medium Ginger knitted vinyl	CF
Avocado knitted vinyl	CG
White knitted vinyl	CW
Black cloth & vinyl	FA
Medium Blue cloth & vinyl	FB
Medium Ginger cloth & vinyl	FF
Avocado cloth & vinyl	FG
Black Mach 1 knitted vinyl	GA
Medium Blue Mach 1 knitted vinyl	GB
Medium Ginger Mach 1 knitted vinyl	GF
Avocado Mach 1 knitted vinyl	GG
White Mach 1 knitted vinyl	GW

Convertible Top Colors

Black or White

1973 Mustang Facts

1973 is significant for it was the last year of the first generation Mustangs. All three Mustang body styles were available but it was the last year for the factory-built convertible. Convertibles would become available again in 1983. 1973 Mustangs were slightly restyled to distinguish them from the 1972s, but in most respects they were unchanged.

The most noticeable change was the redesigned grille. The eggcrate mesh was larger and the turn signal lamps were located within the grille opening. Headlight location remained unchanged; however, the headlight bezel was chrome. Similarly, the taillight bezels were finished in bright metal.

The front bumper, designed to meet the new federal 5 mph standards, was color-keyed to the car's paint. The hood and fender moldings also were color-keyed.

The rear bumper, similar to the 1971-72 unit, was mounted further away from the rear of the car in order to comply with 2½ mph rear standards.

The Decor Group, similar to the 1972 option, used a blacked-out grille with a small Mustang running horse emblem in the center. As with 1972 Mustangs, the optional bodyside tape stripe in black or argent was optional with the Decor Group.

The Mach 1 got a new bodyside tape stripe treatment. A three-quarter vinyl roof was optional on the Mach 1 and all other Sports-Roof models.

The Grande was unchanged, save for colors and vinyl roof treatments.

The non-functional NASA hood was standard equipment on the Mach 1 with the 351 engines and a no-cost option with the 302 engine. The functional hood was optionally available only with the 351 2V engine.

Mustangs with the Tu-tone hood option got the non-functional NASA hood painted either low-gloss black or argent with twist-type hood locks.

Engine availability was unchanged from 1972. The Hurst shifter that came with four-speed equipped Mustangs used a round knob rather than the distinctive Hurst T handle.

1973 convertible

Chapter 10

1974 Mustang II

Production Figures
60F 2dr Hardtop	177,671	69R 3dr Hatchback	
69F 3dr Hatchback	74,799	Mach 1	44,046
60H 2dr Hardtop Ghia	89,477	Total	385,993

Serial Numbers
4R02Y100001

4 — Last digit of model year

R — Assembly plant (F-Dearborn, R-San Jose)

02 — Plate code for Mustang (02-2dr hardtop, 03-3dr hatchback, 04-2dr hardtop Ghia, 05-3dr hatchback Mach 1)

Y — Engine code

100001 — Consecutive unit number

Location
Stamped on plate riveted to driver's dash, visible through the windshield; certification label attached to rear face of driver's door.

Engine Codes
Y — 140 ci 2.3L 2V 4 cyl 88 hp
Z — 171 ci 2.8L 2V V-6 105 hp

1974 Mustang II Prices

	Retail
4 cyl Models	
2dr Hardtop, 60F	$3,134.00
3dr 2+2, 69F	3,328.00
2dr Ghia, 60H	3,480.00
6 cyl Models	
3dr Mach 1, 69R	3,674.00
Extra charge over 2.3 liter 4 cyl	
2.8 liter V-6	229.00
Credit for 2.3 liter substitution on Mach 1	(229.00)
SelectShift Cruise-O-Matic	212.00
Accent Group	151.00
Air conditioner, SelectAire	390.00
Alarm system, anti-theft	75.00
Automatic seatback release	24.00
Axle, Traction-Lok differential	45.00
Battery, heavy-duty	14.00
Belts, color-keyed deluxe	17.00
Bumper guards, front and rear	37.00
Clock, digital quartz crystal	36.00
Console	53.00
Convenience Group:	
Models with Luxury Interior Group	41.00

Mach 1, models w/Rallye Package or Accent Group	21.00
Ghia, models w/Luxury Interior Group in combination with Rallye Package or Mach 1 in combination with Luxury Interior Group	4.00
All other models	57.00
Electric rear window defroster	59.00
Emission equipment, Calif.	19.00
Glass, tinted complete	39.00
Light Group	44.00
Luxury Interior Group	100.00
Maintenance Group	44.00
Mirrors, outside color-keyed remote control	36.00
Molding, rocker panel	14.00
Molding, vinyl insert bodyside	50.00
Paint, glamour	41.00
Pin stripes	14.00
Power front disc brakes	45.00
Power rack and pinion steering	107.00
Protection Group, Mach 1	41.00
All others	47.00
Radio, AM	61.00
Radio, AM/FM monaural	124.00
Radio, AM/FM stereo	222.00
Radio with tape player, AM/FM stereo	346.00
Rallye Package (requires 2.8 V-6):	
Mach 1	150.00
3dr 2+2	200.00
All others	244.00
Rear quarter window, flipper (3dr only)	29.00
Roof, vinyl	83.00
Seat, fold-down rear (std. 3dr)	61.00
Steering wheel, leather-wrapped	30.00
Sunroof, manually operated (2dr only)	149.00
Suspension, competition	37.00
Trim, luggage compartment	28.00
Trim, Picardy velour cloth (Ghia)	62.00
Trim rings (w/styled steel wheels)	32.00
Wheels, four forged aluminum:	
Mach 1, models w/Rallye package, 3dr 2+2 models w/Accent Group	71.00
3dr 2+2 or hardtops with Accent Group	103.00
Ghia	103.00
Hardtop	147.00
Wheels, four styled steel (std. 3dr, models with Rallye Package, and base 2dr with Accent Group)	
Ghia	N/C
Hardtop	44.00
Models having (5) BR78x13 BSW steel-belted radial tires, extra charge for five:	
BR78x13 steel-belted radial WSW	30.00
BR70x13 Wide Oval steel-belted radial B/WL	59.00
CR70x13 Wide Oval steel-belted radial B/WL	77.00

CR70x13 Wide Oval steel-belted radial WSW	65.00
Credit for five:	
B78x13 BSW	(84.00)
B78x13 WSW	(54.00)
Models having (5) BR78x13 WSW steel-belted radial tires, extra charge for five:	
BR70x13 Wide Oval steel-belted radial B/WL	29.00
CR70x13 Wide Oval steel-belted radial B/WL	47.00
CR70x13 Wide Oval steel-belted radial WSW	34.00
Models having (5) BR70x13 B/WL Wide Oval steel-belted radial tires, extra charge for five:	
CR70x13 Wide Oval steel-belted radial B/WL	17.00
CR70x13 Wide Oval steel-belted radial WSW	5.00

1974 Exterior Colors	Code	1974 Exterior Colors	Code
Pearl White	9C	Green Glow	4T
Silver Metallic	1G	Ginger Glow	5J
Bright Red	2B	Tan Glow	5U
Dark Red	2M	**1974 Interior Trim**	**Code**
Light Blue	3B	Black	A
Medium Bright Blue Metallic	3N	Blue	B
Bright Green Gold Metallic	4B	Red	D
		Avocado	G
Medium Lime Yellow	4W	White/Tan	M
Medium Copper Metallic	5M	White/Red	N
		Silver	P
Saddle Bronze Metallic	5T	White/Blue	Q
Medium Yellow Gold	6C	Tan	U

1974 Mustang II Facts

The 1974 Mustang II was a total departure from previous Mustangs. Downsized, it was available only in two body styles—a two-door hardtop and a three-door hatchback. There were no convertibles. It did incorporate many of the first generation Mustang's styling cues such as the long hood/short deck configuration, the side sculpturing and the front grille.

Mechanically, the Mustang II differed in many ways from the first generation Mustangs. The front suspension was redesigned. The front springs were now located between the control arms rather than above the upper A arm. A front subframe was designed to isolate the engine from the rest of the chassis mostly due to the inherent vibration of the standard 2.3L four-cylinder engine. Rack and pinion replaced the recirculating ball steering, and front disc brakes were standard, as were staggered rear shocks. Also standard equipment was a four-speed manual transmission.

In the interior, a more informative dash was used. Tachometer, fuel, alternator and temperature gauges were standard as were non-reclining bucket seats.

Two engines were available. The base 140 ci 2.3L four-cylinder was the first metric American engine. It featured a cross-flow single overhead cam cylinder head, but it pumped out a meager 88 hp—not really enough for a Mustang weighing close to 3,000 pounds.

The only optional engine for 1974 was the German-built 171 ci 2.8L V-6 rated at 105 hp. It was a nice little engine, but again the Mustang II was too heavy for any sort of performance that was reminiscent of the first generation Mustangs.

The Mustang II was available in three models, base in either hardtop or three-door 2+2, the luxury-oriented two-door hardtop Ghia and the performance looking Mach 1 three-door 2+2.

The Ghia moniker replaced the Grande and was the luxury Mustang. Ghia was the name of the Italian design studios that Ford had acquired. Using the Ghia name was intended to lend an air of European exclusivity. The Ghia came with the expected upgrades: deluxe seatbelts, digital quartz clock, the luxury interior group (vinyl seats and door trim, door courtesy lights, 25-oz carpeting, rear ashtray, parking brake boot and sound package), outside color-keyed remote control mirrors, pin stripes, vinyl roof, Picardy velour cloth and the styled steel wheels were a no-cost option.

The Mach 1 came with the larger engine, the 2.8L V-6, and a unique lower bodyside treatment with Mach 1 lettering. However, the Rallye package was required to give the Mach 1 the maximum performance potential. The package consisted of the Traction-Lok differential, CR70x13 wide oval radial B/WL tires, extra cooling package, digital quartz clock, the Competition Suspension (heavy-duty front and rear springs, rear stabilizer bar and adjustable shocks), outside color-keyed remote control mirrors, leather-wrapped steering wheel and styled steel wheels/trim rings.

Other interesting options were a manual sunroof, forged aluminum wheels and an anti-theft alarm system.

1974 three-door Ford Motor Co.

Chapter 11

1975 Mustang II

Production Figures

60F 2dr Hardtop	85,155	69R 3dr Hatchback	
69F 3dr Hatchback	30,038	Mach 1	21,062
60H 2dr Hardtop Ghia	52,320	Total	188,575

Serial Numbers

5F03Y100001

5 — Last digit of model year
F — Assembly plant (F-Dearborn, R-San Jose, T-Metuchen)
03 — Plate code for Mustang (02-2dr hardtop, 03-3dr hatchback, 04-2dr hardtop Ghia, 05-3dr hatchback Mach 1)
Y — Engine code
100001 — Consecutive unit number

Location
Stamped on plate riveted to driver's side of dash; certification label attached to rear face of driver's door.

Engine Codes
Y — 140 ci 2.3L 2V 4 cyl 88 hp
Z — 171 ci 2.8L 2V V-6 105 hp
F — 302 ci 5.0L 2V V-8 140 hp

1975 Mustang II Prices

	Retail
4 cyl 2.3 liter Models	
2dr Hardtop, 60F	$3,529.00
3dr 2+2, 69F	3,818.00
2dr Ghia, 60H	3,938.00
6 cyl 2.8 liter Model	
3dr Mach 1, 69R	4,188.00
2.8 liter V-6 6 cyl (NA with SelectShift)	272.00
302 cid 2V 8 cyl	
Mach 1 (includes SelectShift transmission)	203.00
All others	217.00
Credit for 2.3 liter substitution from base 2.8 V-6	(272.00)
SelectShift Cruise-O-Matic	239.00
Accent Group, exterior	162.00
Air conditioner, SelectAire	417.00
Anti-theft alarm system	76.00
Axle, Traction-Lok differential	46.00
Battery, heavy-duty 53 ampere	14.00
Seatbelts, color-keyed deluxe	17.00
Brakes, power front disc	55.00
Bumper guards, front & rear	35.00
Clock, digital quartz crystal	40.00

Console	63.00
Convenience Group (depending on model)	7.00-70.00
Defroster, electric rear window	63.00
Emission equipment, Calif.	41.00
Fuel tank, extended range (std. 302 V-8)	19.00
Glass, tinted complete	41.00
Light Group	35.00
Light, fuel monitor warning	19.00
Lock Group, security	15.00
Luxury Interior Group, Hardtop	106.00
3dr 2+2 and Mach 1	89.00
Luxury Group, Ghia Silver	162.00
Maintenance Group	48.00
Mirrors, outside dual color-keyed	39.00
Molding, rocker panel	19.00
Molding, color-keyed vinyl insert bodyside	51.00
Moonroof, silver glass	454.00
Paint, glamour	49.00
Protection Group, Mach 1	20.00
All others	29.00
Radio, AM	65.00
Radio, AM/FM monaural	136.00
Radio, AM/FM stereo	225.00
Radio with tape player, AM/FM stereo	347.00
Rallye Package, Mach 1	141.00
3dr 2+2	195.00
All others	282.00
Roof, vinyl	83.00
Roof, vinyl half, Ghia	N/C
Seat, fold down rear (std. 3dr)	66.00
Steering, power rack & pinion	117.00
Steering wheel, leather-wrapped	32.00
Stripes, pin	24.00
Sunroof, manually operated	210.00
Suspension, Competition, Ghia & Accent Group	43.00
Mach 1	25.00
All others	55.00
Trim, velour cloth, Ghia	88.00
Trim rings	35.00
Wheels, four cast aluminum spoke	
Mach 1, Rallye Package, 3dr 2+2 with Exterior Accent Group	78.00
3dr 2+2 or Hardtop with Exterior Accent Group	113.00
Ghia	113.00
Hardtop without Exterior Accent Group	158.00
Wheels, four styled steel, Ghia	N/C
Hardtop	45.00
Windows, pivoting rear quarter(3dr)	33.00
Credit for deletion of color-keyed deluxe seatbelts	(17.00)
Credit for deletion of digital clock	(40.00)
Credit for deletion of tinted glass-complete	(41.00)
Typical tire upgrade	30.00-105.00

1975 Exterior Colors	Code
Black	1C
Silver Metallic	1G
Bright Red	2B
Dark Red	2M
Bright Blue Metallic	3E
Silver Blue Glow	3M
Pastel Blue	3Q
Green Glow	4T
Dark Yellow Green Metallic	4V
Light Green	47
Medium Copper Metallic	5M
Dark Brown Metallic	5Q
Tan Glow	5U
Bright Yellow	6E
Polar White	9D

1975 Interior Trim	Code
Black	A
Blue	B
Red	D
Green	G
Cranberry	H
White/Red	N
White/Blue	Q
Tan	U
White/Tan	4
White/Green	5

1975 Mustang Facts

The 1975 Mustang II was hardly changed. The grille got a larger eggcrate-type mesh, which was now practically flush with the grille opening, and the Ghia model came with opera windows to enhance its luxury image.

The 2.3L four-cylinder engine was still standard equipment, with the 2.8L V-6 optional, again available only with a four-speed manual. The 302 ci V-8 rated at 140 hp became optional on all Mustangs.

The Ghia Luxury Group, optional on the Ghia, included silver metallic paint, a half vinyl roof in Silver Normande grain, full-length bodyside tape stripes, stand-up hood ornament, media velour cloth trim in cranberry, console, flocked headlining and sun visor.

There was also the regular Luxury Interior Group (standard on the Ghia) which included a choice of vinyl or cloth and vinyl seat trim, deluxe door and rear seat quarter trim, door courtesy lights, color-keyed deluxe belts on hardtops, shag carpeting, rear ashtray, parking brake boot and, as Ford called it, a super sound package.

Two sunroofs were available, the silver glass moonroof and the regular version, both manually controlled.

The Rallye Package, available only with the 2.8 V-6 or the 302 V-8, tightened the chassis up a bit for better handling. It included the Traction-Lok differential, 195/70 BWL tires, extra cooling package, exhaust with bright tips, digital clock (hardtops), the Competition Suspension, remote control outside color-keyed mirrors, leather-wrapped steering wheel and styled steel wheels with trim rings.

The Competition Suspension, available by itself, included heavy-duty springs, Gabriel adjustable shocks, a rear stabilizer bar and 195/70x13 B/WL tires.

From the performance point of view, the availability of the 302 ci V-8 helped give the Mustang II a much needed shot in the arm. Available only with the three-speed automatic transmission, mandatory options were power brakes and steering. California-bound

302s got catalytic converters and all Mustang engines benefited from electronic ignition.

Steel-belted radial tires were now standard equipment.

New wheels became available. These were a cast aluminum spoke-type wheel. The styled steel and forged aluminum wheels were also available.

Late in the model year, an MPG version of the Mustang II was made available. Using the 2.3L four-cylinder engine and a lower numerical rear axle ratio, 3.18:1 vs 3.40:1, the MPG Mustang was designed to deliver better mileage. New lows were reached in terms of acceleration.

1975 two-door Ghia Ford Motor Co.

Chapter 12

1976 Mustang II

Production Figures
60F 2dr Hardtop	78,508	69R 3dr Hatchback	
69F 3dr Hatchback	62,312	Mach 1	9,232
60H 2dr Hardtop Ghia	37,515	Total	187,567

Serial Numbers
6F03Y100001

6 — Last digit of model year

F — Assembly plant (F-Dearborn, R-San Jose)

03 — Plate code for Mustang (02-2dr hardtop, 03-3dr hatchback, 04-2dr hardtop Ghia, 05-3dr hatchback Mach 1)

Y — Engine code

100001 — Consecutive unit number

Location
Stamped on plate riveted on driver's side of dash; certification label attached to rear face of driver's door.

Engine Codes
Y — 140 ci 2.3L 2V 4 cyl 88 hp

Z — 171 ci 2.8L 2V V-6 105 hp

F — 302 ci 5.0L 2V V-8 140 hp

1976 Mustang II Prices

	Retail
4 cyl 2.3 liter Models	
MPG 2dr Hardtop, 60F	$3,525.00
MPG 3dr 2+2, 69F	3,781.00
MPG 2dr Ghia, 60H	3,859.00
6 cyl 2.8 liter Models	
2dr Hardtop, 60F	3,791.00
3dr 2+2, 69F	4,047.00
2dr Ghia, 60H	4,125.00
3dr Mach 1, 69R	4,209.00
8 cyl 302 cid Models	
2dr Hardtop, 60F	3,737.00
3dr 2+2, 69F	3,992.00
2dr Ghia, 60H	4,071.00
3dr Mach 1, 69R	4,154.00
Credit for 2.3 liter substitution on Mach 1	(272.00)
4-speed manual heavy-duty transmission (required with 302 cid unless Cruise-O-Matic is ordered)	37.00
SelectShift Cruise-O-Matic	239.00
Accent Group, exterior	169.00
Air conditioner, SelectAire (requires power steering)	420.00
Anti-theft alarm system	83.00

Axle, optional ratio	13.00
Axle, Traction-Lok differential (requires power front disc brakes)	48.00
Battery, heavy-duty (std. 2.8 & 302 cid)	14.00
Seatbelts, deluxe color-keyed	17.00
Black Midnight option (Mach 1 only)	83.00
Bracket, front license plate	N/C
Brakes, power front disc	54.00
Bumper guards, front & rear	34.00
Clock, digital quartz crystal	40.00
Clock, electric (NA w/Rallye Package)	17.00
Cobra II Package	325.00
Cobra II Modification Package	287.00
Console	71.00
Convenience Group	35.00
Defroster, electric rear window	70.00
Emission equipment, Calif.	49.00
Fuel tank, extended range	24.00
Ghia Luxury Group	177.00
Glass, tinted complete	46.00
Heater, engine block	17.00
Horn, dual note	6.00
Light Group, models with sunroof or moonroof	28.00
All others	41.00
Light, fuel monitor warning	18.00
Lock Group, security	16.00
Luggage rack, deck lid	51.00
Luxury Interior Group	117.00
Mirrors, outside dual color-keyed	42.00
Molding, color-keyed vinyl insert bodyside	60.00
Molding, rocker panel	19.00
Moonroof, glass	470.00
Paint, glamour	54.00
Paint/tape, Tu-tone	84.00
Protection Group, Mach 1 & models w/Black Midnight option	36.00
All others	43.00
Radio, AM	71.00
Radio with tape player, AM	192.00
Radio, AM/FM monaural	128.00
Radio, AM/FM stereo	299.00
Rallye Package, Mach 1 & w/Cobra II	163.00
All others	237.00
Roof, vinyl	86.00
Roof, half vinyl, Ghia only	N/C
Seat, fold-down rear (std. 3dr)	72.00
Stallion Group, Hardtop & 2+2	72.00
Steering, power rack & pinion (required with 302 cid & 2.8 liter w/air conditioning; power front disc brakes also required)	117.00
Steering wheel, leather-wrapped	33.00
Stripes, pin	27.00

Sunroof, manually operated	230.00
Suspension, Competition	29.00
Trim, velour cloth, Ghia only	99.00
Trim rings	35.00
Wheels, four cast aluminum spoke	
Mach 1, Rallye Package or Cobra II	96.00
Ghia, Stallion Group or Exterior Accent Group	131.00
All others	182.00
Wheels, four forged aluminum	
Mach 1, Rallye Package or Cobra II	96.00
Ghia, Stallion Group or Exterior Accent Group	131.00
All others	182.00
Wheels, four styled steel	
Hardtop & 2+2	51.00
Ghia	N/C
Windows, pivoting rear	33.00
Fleet options:	
Light, luggage compartment	4.00
Mirror, lefthand color-keyed	14.00
Tinted glass, windshield	24.00
Tire upgrade	33.00-208.00

1976 Exterior Colors

Color	Code
Black	1C
Silver Metallic	1G
Bright Red	2B
Dark Red	2M
Bright Blue Metallic	3E
Silver Blue Glow	3M
Medium Ivy Bronze Metallic	4T
Dark Yellow Green Metallic	4V
Light Green	47
Medium Chestnut Metallic	5M
Dark Brown Metallic	5Q
Tan Glow	5U
Bright Yellow	6E
Polar White	9D

Tu-tone Exterior Color Combinations

Body color/accent color	Code
Cream/Medium Gold Metallic	6P/6V
White/Bright Red	9D/2R
White/Bright Blue Metallic	9D/3E

1976 Interior Trim

Trim	Code
Black	A
Blue	B
Red	D
Cranberry	H
Aqua	K
Tan	U
White/Red	N
White/Blue	Q
White/Tan	4
White/Black	W

1976 Mustang II Facts

The 1976 Mustang II continued with few changes. Most options and packages were carried over from 1975.

Base Mustang models with the 2.3 were all classified as MPG models, stressing their economy features.

The 2.8L V-6 was now available with an automatic transmission, at extra cost.

A wide ratio four-speed manual became available on the 302 V-8.

The Stallion Group, coinciding with similar offerings on the Maverick and Pinto, was strictly a cosmetic package. It included black greenhouse moldings and wiper arms, black grille (horse grille emblem deleted), black rocker panels, lower fenders, lower doors, lower front and rear bumpers, and black lower quarter panels. Also included were four styled steel wheels, bright lower bodyside moldings and Stallion decals. The wheel lip moldings were deleted.

The Cobra II modification package was installed by Motortown Corporation. It included: a front air spoiler, rear deck lid spoiler and simulated hood scoop; quarter window louvers with "snake" emblems; accent stripes on front spoiler, hood, roof, rear deck, rear deck spoiler and lower bodyside panels; "snake" emblems on the fenders and wheel centers; and a "snake" emblem on the blacked-out grille.

Paint schemes were white with blue stripes or black and gold stripes. All Mustang engines were available, but it was the 302 V-8 that helped fullfill the promises made by the Cobra II package.

1976 three-door Stallion Ford Motor Co.

Chapter 13

1977 Mustang II

Production Figures
60F 2dr Hardtop	67,783	69R 3dr Hatchback	
69F 3dr Hatchback	49,161	Mach 1	6,719
60H 2dr Hardtop Ghia	29,510	Total	153,173

Serial Numbers
7F03Y100001

7 — Last digit of model year

F — Assembly plant (F-Dearborn, R-San Jose)

03 — Plate code for Mustang (02-2dr Hardtop, 03-3dr Hatchback, 04-2dr Hardtop Ghia, 05-3dr Hatchback Mach 1)

Y — Engine code

100001 — Consecutive unit number

Location

Stamped on plate riveted on driver's side of dash, visible through the windshield; certification label attached on rear face of driver's door.

Engine Codes

Y — 140 ci 2.3L 2V 4 cyl 92 hp

Z — 171 ci 2.8L 2V V-6 103 hp

F — 302 ci 5.0L 2V V-8 134 hp

1977 Mustang II Prices

	Retail
4 cyl 2.3 liter Models	
2dr Hardtop, 60F	$3,678.00
3dr 2+2, 69F	3,877.00
2dr Ghia, 60H	4,096.00
6 cyl 2.8 liter Models	
3dr Mach 1, 69R	4,332.00
Credit for 2.3 liter substitution on Mach 1	(306.00)
2.8 liter 6 cyl	306.00
302 cid 8 cyl, Mach 1	(12.00)
All others	294.00
SelectShift Cruise-O-Matic (required with 302 cid)	248.00
Four-speed manual (NA 302 cid)	N/C
Appearance Decor Group (NA Mach 1, Ghia, Cobra II)	
Hardtop	151.00
2+2	106.00
Accent Group, exterior, Hardtop	211.00
Air conditioner, SelectAire	443.00
Seatbelts, color-keyed deluxe	17.00
Brakes, power front disc	57.00
Bumper guards, front & rear	35.00

Clock, digital quartz crystal	42.00
Cloth & vinyl bucket seats, Hardtop & 2+2	12.00
Cobra II Package	689.00
Console	75.00
Convenience Group, 3dr (NA Cobra II & T-roof)	65.00
All others	33.00
Defroster, electric rear window	70.00
Fold-down rear seat (std. 3dr)	84.00
Ghia Sports Group (Ghia only)	398.00
Luxury Interior Group	147.00
Glass, tinted complete	49.00
T-roof convertible option (2+2 & Mach 1)	
2+2 with Cobra II	587.00
All others	629.00
Light Group, w/manually operated sunroof,	
flip-up open air roof & T-roof	37.00
All others	42.00
Luggage rack, deck lid	52.00
Mirrors, dual sport	47.00
Moldings, color-keyed vinyl insert bodyside	63.00
Moldings, rocker panel	20.00
Bodyside molding color-keyed to exterior paint	N/C
Stripes, pin	28.00
Protection Group	
Models w/front license plate bracket	
Mach 1 & Cobra II	27.00
All others	34.00
Models without bracket	
Mach 1 & Cobra II	23.00
All others	30.00
Radio, AM	65.00
Radio with stereo tape player, AM	192.00
Radio, AM/FM monaural	120.00
Radio, AM/FM stereo, models	
with Deluxe Equipment Group	41.00
All others	161.00
Radio with tape player, AM/FM stereo	
Models with Deluxe Equipment Group	108.00
All others	229.00
Rallye Appearance Package, 2+2	157.00
Rallye Package, Mach 1, Cobra II, Exterior Accent Group	
& Rallye Appearance Package	54.00
All others	101.00
Seat, 4-way manual driver's	30.00
Sports Performance Package, includes 302 cid V-8	
heavy-duty 4-speed manual transmission,	
power steering, power brakes and 195/70R WSW	
(RWL on 2+2 & Mach 1) tires	
Hardtop with Exterior Accent Group	649.00
Hardtop without Exterior Accent Group	686.00
2+2 model with Cobra II package	516.00
2+2 without Cobra II package	649.00

Ghia	593.00
Mach 1	210.00
Discount, Mach 1	(122.00)
All others	(135.00)
Steering, power rack & pinion	125.00
Steering wheel, leather-wrapped sport	
2+2 & Mach 1	35.00
All others	51.00
Trim, media velour cloth, Ghia only	102.00
Roof, flip-up open air, 2dr only	145.00
Sunroof, manually operated, 2dr only	237.00
Roof, full vinyl, Hardtop only	88.00
Paint, metallic glow	57.00
Trim rings, 2+2 only	36.00
Wheel covers, wire, Hardtop	82.00
2+2	40.00
Ghia	63.00
Mach 1, Cobra II, Exterior Accent Group, Appearance Decor Group or Rallye Appearance Package	4.00

Wheels, four lacy spoke aluminum (NA Ghia Sports Group)

Hardtop	204.00
2+2	161.00
Ghia	184.00
Mach 1, Cobra II, Exterior Accent Group, Appearance Decor Group or Rallye Appearance Package	125.00

Wheels, four forged aluminum (NA Ghia Sports Group)

Hardtop	204.00
2+2	161.00
Ghia	184.00
Mach 1, Cobra II, Exterior Accent Group, Appearance Decor Group or Rallye Appearance Package	125.00

Wheels, four white lacy spoke aluminum (NA Ghia Sports Group or Rallye Appearance Package)

Hardtop	252.00
2+2	210.00
Ghia	233.00
Mach 1, Cobra II, Exterior Accent Group or Appearance Decor Group	173.00

Wheels with trim rings, four styled steel (std. with Mach 1 and 2+2 [less trim rings]), Exterior Accent Group, Appearance Decor Group and Rallye Appearance Package)

Hardtop	78.00
Ghia	59.00
Spoiler, front, 2+2 and Mach 1	N/C
Battery, heavy-duty	16.00
Bracket, front license plate	N/C
High altitude emission equipment	22.00
Emission equipment, Calif.	69.00
Tires, typical upgrade	18.00–185.00

Limited production options	
Heater, engine block immersion	18.00
Light, luggage compartment	4.00
Mirror, inside day/night	7.00
Fleet options	
Mirror, lefthand sport	14.00
Tinted glass, windshield	24.00

1977 Exterior Colors

Colors	Code
Black	1C
Bright Red	2R
Dark Brown Metallic	5Q
Bright Yellow	6E
Cream	6P
Golden Glow	6V
Bright Aqua Glow	7H
Light Aqua Metallic	7Q
Medium Emerald Glow	7S
Orange	8G
Tan	8H
Bright Saddle Metallic	8K
Polar White	9D

Tu-tone Exterior Color Combinations*

Body color/accent color	Code
Cream/Medium Gold Metallic	6P/6V
White/Bright Red	9D/2R
White/Light Aqua Metallic	9D/7Q

*Before 6-27-77; after, see 1978

1977 Interior Trim*

	Code
Black	A
Red	D
Aqua	K
Chamois	T
Cream	V
White/Red	N
White/Black	W
White/Chamois	2
White/Emerald	5
White/Aqua	7
White/Gold	8

*Before 6-27-77; after, see 1978

1977 Mustang II Facts

In 1977, two-door models got a new horizontal grille while all SportsRoof Mustangs came with a blacked-out grille.

SportsRoofs were also upgraded by the addition of a standard sport steering wheel, styled steel wheels with raised white letters, bias belted tires and brushed aluminum instrument panel appliques. Hardtops came with pecan-colored woodgrain appliques.

A new interior option was the four-way adjustable seats. These were adjustable for height in addition to forward and back. The seatback was still non-adjustable.

The front spoiler used on the Cobra II was available on all SportsRoofs as a no-cost option.

The moonroof was deleted but the regular manually-controlled sunroof remained. A flip-up open air roof was also available. It either flipped up in the rear, or could be removed completely.

1977 was the first year for the T-roof option, consisting of removable roof panels, but it was only available on the three-door models.

The Appearance Decor Group, not available on the Ghia or Mach 1, included lower body Tu-tone paint treatment, pin stripes, four styled steel wheels with trim rings, all vinyl or cloth and vinyl seat trim, and brushed aluminum instrument panel appliques. Wheel lip moldings were deleted.

Available only on hardtops was the Exterior Accent Group, consisting of pin stripes, wide color-keyed vinyl insert bodyside moldings, WSW tires, dual sport mirrors and four styled steel wheels with trim rings.

Complementing the Rallye Package was a Rallye Appearance Package, available only on the three-door. It was primarily a blacked-out treatment.

The Sports Performance Package consisted of the 302 ci V-8 engine, four-speed manual transmission, power steering and brakes, and 195/70R WSW tires.

Available only on the Ghia was the unique Ghia Sports Group. It consisted of black or tan exterior paint, chamois or black half-vinyl roof, vinyl insert bodyside moldings, pin stripes, luggage rack with black or chamois hold-down straps with bright buckles, lacy spoke aluminum wheels with chamois painted spokes and blacked-out grille. In the interior, the Ghia seat trim was finished in chamois with black upper straps on the seatbacks, black engine-turned applique inserts on the instrument panel, console tray-door, trim panel inserts, console, leather-wrapped steering wheel, black shift lever with manual transmission and black parking brake handle.

The Ghia Silver Luxury Group was deleted for 1977.

Performance image duties were still handled by the Mach 1 and SportsRoof with the Cobra II package. Two additional stripe colors, red and green, were available on white Cobra IIs. All 1977 and later Cobra IIs were built by Ford.

California-bound 2.8L V-6s and 302 ci V-8s were equipped with a new Variable-Venturi carburetor design.

The cast aluminum spoke wheels were deleted. Lacy spoke wheels in natural or painted white became options.

1977 Cobra II Joy Jacobs Ford Motor Co.

Chapter 14

1978 Mustang II

Production Figures

60F 2dr Hardtop	81,304	69R 3dr Hatchback	
69F 3dr Hatchback	68,408	Mach 1	7,968
60H 2dr Hardtop Ghia	34,730	Total	192,410

Serial Numbers
8F03Y100001
8 — Last digit of model year
F — Assembly plant (F-Dearborn, R-San Jose)
03 — Plate code for Mustang (02-2dr hardtop, 03-3dr hatchback, 04-2dr hardtop Ghia, 05-3dr hatchback Mach 1)
Y — Engine code
100001 — Consecutive unit number

Location
Stamped on plate riveted on driver's side of dash, visible through the windshield; certification label attached on rear face of driver's door.

Engine Codes
Y — 140 ci 2.3L 2V 4 cyl 88 hp
Z — 171 ci 2.8L 2V V-6 90 hp
F — 302 ci 5.0L 2V V-8 139 hp

1978 Mustang II Prices   Retail

2.3 liter 4 cyl Models	
2dr Hardtop	$3,824.00
3dr 2+2	4,088.00
2dr Ghia	4,242.00
2.8 liter 6 cyl Models	
3dr Mach 1	4,523.00
Credit for 2.3 liter substitution from 2.8 liter, Mach 1	(237.00)
2.8 liter 2V 6 cyl (Variable Venturi in Calif.)	237.00
5.0 liter (302 cid) 8 cyl	
Mach 1	148.00
All others	386.00
SelectShift Cruise-O-Matic	292.00
Accent Group, Exterior	
Hardtop	245.00
2+2	163.00
Air conditioner, SelectAire	469.00
Appearance Decor Group	
Hardtop	167.00
2+2	128.00
Seatbelts, color-keyed deluxe	19.00

Bodyside protection, lower	30.00
Bracket, front license plate	N/C
Brakes, power front disc	66.00
Bumper guards, front and rear	39.00
Clock, digital quartz crystal	46.00
Cobra II Package	
Models with 2.3 or 2.8 liter engine	701.00
Models with 5.0 liter engine	724.00
Console	75.00
Convenience Group	
3dr, except Cobra II or T-roof	81.00
All others	34.00
Defroster, electric rear window	80.00
Emission equipment, Calif.	69.00
Emission equipment, high altitude	33.00
Fashion Accessory Package	219.00
Fold-down rear seat	90.00
Ghia Sports Group	386.00
Glass, tinted complete	54.00
Heater, engine block immersion	12.00
Illuminated entry system	49.00
King Cobra option	1,277.00
Light Group	
Models with flip-up roof or T-roof	40.00
All others	52.00
Luxury Interior Group	
Hardtop	167.00
2+2 and Mach 1	161.00
Mirror, lefthand illuminated visor vanity	34.00
Mirrors, dual sport	49.00
Moldings, color-keyed vinyl insert bodyside	66.00
Moldings, bodyside color-keyed to exterior paint	N/C
Moldings, rocker panel	22.00
Paint, metallic glow	40.00
Protection Group	
Models with front license plate bracket	
Mach 1	28.00
All others	36.00
Models without front license plate bracket	
Mach 1	24.00
All others	33.00
Radio flexibility option	90.00
Radio, AM	72.00
Radio with 8-track stereo tape player, AM	192.00
Radio, AM/FM monaural	120.00
Radio, AM/FM stereo	161.00
Radio with cassette player, AM/FM stereo	229.00
Radio with 8-track tape player, AM/FM stereo	229.00
Rallye Appearance Package	163.00
Rallye Package	
Mach 1, Exterior Accent Group,	
Rallye Appearance Package	43.00

All others	93.00
Roof, flip-up open air	167.00
Roof, full vinyl	99.00
Seat, 4-way manual driver's	33.00
Spoiler, front	8.00
Steering, power rack & pinion	134.00
Steering wheel, leather-wrapped sport	
2+2, Mach 1	34.00
All others	49.00
Stripes, pin	30.00
T-roof convertible option	
2+2 without Cobra II, Mach 1	689.00
2+2 with Cobra II	647.00
Trim, cloth & vinyl (Ashton cloth)	12.00
Trim, Willshire cloth	100.00
Trim rings	39.00
Wheel covers, wire	
Hardtop	96.00
2+2	45.00
Ghia	77.00
Mach 1, Cobra II, Exterior Accent Group, Appearance Decor Group, Rallye Appearance Package	12.00
Wheels, four lacy spoke aluminum	
Hardtop	276.00
2+2	224.00
Ghia	257.00
Mach 1, Cobra II, Exterior Accent Group, Appearance Decor Group, Rallye Appearance Package	186.00
Wheels, four forged aluminum	
Hardtop	276.00
2+2	224.00
Ghia	257.00
Mach 1, Cobra II, Exterior Accent Group, Appearance Decor Group, Rallye Appearance Package	186.00
Wheels, four white lacy spoke aluminum	
Hardtop	289.00
2+2	237.00
Ghia	270.00
Mach 1, Cobra II, Exterior Accent Group, Appearance Decor Group, Rallye Appearance Package	199.00
Wheels with trim rings, four white styled steel	
Hardtop	90.00
Ghia	71.00
Wheels, four white painted forged aluminum	
Hardtop	289.00
2+2	237.00
Ghia	270.00
Mach 1, Cobra II, Exterior Accent Group, Appearance Decor Group, Rallye Appearance Package	199.00

Typical tire upgrade	23.00–202.00
Limited Production options	
Light, luggage compartment	4.00
Mirror, inside day/night	7.00
Mirror, lefthand color-keyed sport	16.00
Tinted glass, windshield	25.00

1978 Exterior Colors

Color	Code
Black	1C
Silver Metallic	1G
Bright Red	2R
Dark Midnight Blue	3A
Dark Jade Metallic	46
Medium Chestnut Metallic	5M
Dark Brown Metallic	5Q
Bright Yellow	6E
Aqua Glow	7H
Aqua Metallic	7Q
Chamois Glow	8W
Light Chamois	83
Tangerine	85
Polar White	9D

Tu-tone Exterior Color Combinations

Body color/accent

color	Code
Silver Metallic/Black	1G/1C
Silver Metallic/Bright Red	1G/2R
Bright Red/Black	2R/1C
Bright Red/White	2R/9D
Dark Jade Metallic/White	46/9D
Bright Yellow/Black	6E/1C
Bright Aqua Metallic/White	7H/9D
Light Aqua Metallic/White	7Q/9D
Light Chamois/Medium Chestnut Metallic	83/5M
Tangerine/White	85/9D
White/Black	9D/1C
White/Bright Red	9D/2R
White/Aqua Metallic	9D/7Q

1978 Interior Trim

Trim	Code
Black	A
Red	D
Tangerine	J
Aqua	K
White/Red	N
Chamois	T
White/Black	W
White/Chamois	2
White/Aqua	7
White/Gold	8

1978 Mustang II Facts

1978 was the last year for the Mustang II, the Mach 1 model and the Cobra II option package.

In the interior, the most noticeable change was the use of two rear seat cushions replacing the previous full-length seat.

Mechanically, the 302 ci V-8 came with either a two-barrel or the Variable Venturi carburetor. The optional power steering was enhanced through the use of variable ratio.

The Cobra II got a new tape stripe treatment, and black rear window louvers, similar to the Sport Slats of 1969-70 Mustangs, were made part of the package.

The Fashion Accessory Package, specifically designed to appeal to women, consisted of Fresno cloth seat inserts, a driver's side illuminated sun visor mirror (useful in slow traffic), a four-way adjustable driver's seat, coin tray, map pockets for maps and other things, an illuminated entry system and exterior tape stripes.

The optional styled steel wheels with trim rings were available in white only. The forged aluminum wheels were also available in white, in addition to natural aluminum.

The most expensive option seen on the Mustang since the Boss 429 engine was the King Cobra option, at $1,277.00. This was a special cosmetic treatment/equipment package for the Sports-Roof. It included a special front air dam, the Cobra II's hood scoop and rear spoiler, wheels similar to those found on the Pontiac Trans AM, special striping and hood decal, and King Cobra identification. Mechanically, the 302 ci V-8 was standard equipment as was power steering, power brakes, heavy-duty springs, adjustable shocks, rear stabilizer bar and spoke wheels.

1978 Cobra II Ford Motor Co.

1978 King Cobra

Chapter 15

1979 Mustang

Production Figures

66B 2dr Sedan	156,666	61H 3dr Hatchback	
61R 3dr Hatchback	120,535	Ghia	36,384
66H 2dr Sedan Ghia	56,351	Total	369,936

Serial Numbers

9F02Y100001

9 — Last digit of model year
F — Assembly plant (F-Dearborn, R-San Jose)
02 — Plate code for Mustang (02-2dr sedan, 03-3dr hatchback, 04-2dr Ghia, 05-3dr Ghia)
Y — Engine code
100001 — Consecutive unit number

Location

Stamped on riveted plate on driver's side of dash, visible through the windshield; certification label attached on rear face of driver's door.

Engine Codes

Y — 2.3 liter 2V 4 cyl 88 hp
W — 2.3 liter 2V 4 cyl 132 hp (Turbocharged)
Z — 2.8 liter 2V V-6 109 hp
T — 3.3 liter 1V 6 cyl 85 hp
F — 5.0 liter 2V V-8 140 hp

1979 Mustang Prices

	Retail
2.3 liter 4 cyl Models	
2dr Sedan	$4,494.00
3dr Sedan	4,828.00
2dr Ghia	5,064.00
3dr Ghia	5,216.00
2.8 liter 6 cyl engine	273.00
2.3 liter 4 cyl turbocharged	542.00
3.3 liter 6 cyl	241.00
5.0 liter, with Cobra Package	N/C
All others	514.00
SelectShift automatic	307.00
Accent Group, exterior	72.00
Accent Group, interior	
2dr	120.00
3dr	108.00
Air conditioner, SelectAire	484.00
Battery, heavy-duty	18.00

Seatbelts, color-keyed deluxe	20.00
Bodyside protection, lower	30.00
Bracket, front license plate	N/C
Brakes, power front disc	70.00
Cobra Package	1,173.00
Cobra hood graphics	78.00
Console	140.00
Deflectors, mud & stone	23.00
Defroster, electric rear window	84.00
Emission system, Calif.	76.00
Emission system, high altitude	33.00
Exhaust, sport-tuned	34.00
Glass, tinted complete	59.00
Light Group, without flip-up open air roof	37.00
with open air roof	25.00
Lock Group, power	99.00
Mirror, lefthand remote-control	18.00
Mirrors, dual remote	52.00
Moldings, narrow vinyl insert bodyside	39.00
Moldings, rocker panel	24.00
Moldings, wide bodyside	66.00
Paint, metallic glow	41.00
Paint, lower Tu-tone	78.00
Protection Group, with front license plate bracket	36.00
without bracket	33.00
Radio flexibility option	90.00
Radio, AM	84.00
Radio, AM, digital clock	131.00
Radio, AM/FM monaural	133.00
Radio, AM/FM stereo	188.00
Radio, AM/FM stereo with cassette tape	255.00
Radio, AM/FM stereo with 8-track tape	255.00
Roof, flip-up open air	199.00
Roof, full vinyl	102.00
Seat, 4-way manual driver's	35.00
Sound system, premium	67.00
Speakers, dual rear seat	42.00
Speed control, fingertip, 2dr without Sport option	116.00
All others	104.00
Sport option	175.00
Steering, power variable ratio	141.00
Steering wheel, leather-wrapped sport	
2dr without Sport option	53.00
All others	41.00
Steering wheel, tilt	
2dr without Sport option	81.00
All others	69.00
Stripes, pin (bodyside and deck lid)	30.00
Suspension, handling	33.00
Trim, leather low-back bucket seat	282.00
Trim, cloth & vinyl seat	
Sedan	20.00

Ghia	42.00
Trim, accent cloth and vinyl seat	29.00
Wheel covers, four turbine	
3dr or with Sport option	10.00
All others	39.00
Wheel covers, four wire	
3dr or with Sport option	70.00
Ghia	60.00
All others	99.00
Wheels, four cast aluminum	
3dr or with Sport option	260.00
Ghia	251.00
All others	289.00
Wheels, four forged metric aluminum	
3dr or with Sport option	269.00
Ghia	259.00
All others	298.00
Wheels with trim rings, four styled steel	
3dr or with Sport option	65.00
Ghia	55.00
All others	94.00
Windshield wipers, interval	35.00
Wiper/washer, rear window	63.00
Limited Production options	
Floor mats, front (color-keyed)	18.00
Heater, engine block immersion	13.00
Light, luggage compartment	5.00
Tinted glass, windshield	25.00
Tires, Models having four B78x13 bias ply BSW as standard equipment, extra charge for:	
B78x13 bias WSW	43.00
C78x13 bias BSW	25.00
C78x13 bias WSW	69.00
B78x14 bias WSW	66.00
C78x14 bias BSW	48.00
BR78x14 BSW	124.00
BR78x14 WSW	167.00
CR78x14 WSW	192.00
CR78x14 RWL	209.00
TRX 190/65Rx390 BSW Michelin*	241.00
Models having four BR78x14 BSW tires as standard equipment, extra charge for:	
BR78x14 WSW	43.00
CR78x14 WSW	69.00
CR78x14 RWL	86.00
TRX 190/65Rx390 BSW Michelin*	117.00
*Requires forged metric aluminum wheels	

1979 Exterior Colors	**Code**	**1979 Exterior Colors**	**Code**
Black	1C	Red Glow	2H
Silver Metallic	1G	Bright Red	2P

1979 Exterior Colors

Color	Code
Light Medium Blue	3F
Medium Blue Glow	3H
Bright Blue	3J
Dark Jade Metallic	46
Medium Chestnut Metallic	5M
Medium Vaquero Gold	5W
Bright Yellow	64
Light Chamois	83
Tangerine	85
Medium Grey Metallic	1P
Polar White	9D

Tu-tone Exterior Color Combinations

Body color/accent color	Code
All/Black	/1C

Tu-tone Exterior Color Combinations

	Code
Silver Metallic/ Medium Grey Metallic	1G/1P
Light Medium Blue/ Bright Blue	3F/3J

1979 Interior Trim

Trim	Code
Black	A
Red	D
Wedgewood Blue	B
Chamois	T
White/Red	N
White/Black	W
White/Chamois	2
White/Blue	Q

1979 Mustang Facts

The Mustang went through its third major change in 1979. Retaining the basic dimensions of the Mustang II, the Mustang's wheelbase was increased to 100.4 inches (from 96.2), which translated into a roomier passenger compartment. The 1979 Mustang was a totally new design with no styling references to the original Mustang or Mustang II. Two body styles were available, a two-door sedan and the three-door hatchback.

Sharing many components from the Fairmont/Zephyr platform, the 1979 Mustang was the first Mustang to use a strut-type front suspension—with a slight variation. Whereas the typical strut suspension has the front coil spring mounted around the strut, in the Mustang the coil spring was located between the lower control arm and chassis. The rear 6.75 inch axle was located by a four-bar link arrangement as the Mustang used coil springs in the rear for the first time. Steering was a carryover and brakes were front discs/rear drums.

Engine availability was similar to the Mustang II's. The 2.3 liter four-cylinder was standard equipment, the German made 2.8 liter V-6 optional and the 5.0 liter V-8 was still available as the only V-8 option. For the first time, the 302 was fitted with a single-belt (serpentine) accessory drive system.

An unreliable turbocharged version of the 2.3 liter four-cylinder was optional as well, generating 132 hp. A lot of hoopla was made about this engine, but Ford was forced to withdraw it for a variety of mechanical ills after the 1980 model year.

The old 200 ci inline six (3.3L) replaced the German V-6 midway through the model year.

An effort was made to improve the Mustang's handling. The optional handling suspension came with a 0.50 inch rear stabilizer bar (0.56 inch with the 2.8 V-6). The 0.50 inch bar was also standard with the 302 ci engine. More significant was the suspension package that was included when the Michelin 190/65R 390 TRX tires and TR 390 forged aluminum wheels were ordered. Shocks, springs and stabilizer bars were tuned to the TRX wheels/tires to provide excellent handling, especially on smooth surfaces. As good as the tires were, the 302's torque could easily overwhelm them resulting in wheel spin and lots of wheel hop. The wheel hop problem, due more to the rear suspension's design, wasn't really effectively solved until 1984 with the introduction of the Quadra-Shock rear on the SVO and later on the 5.0L GTs.

The Ghia model still stressed luxury with usual interior upgrades and Ghia emblems. The performance-oriented option was the Cobra package, available only on the three-door hatchback. Standard engine was the 2.3L Turbocharged four-cylinder, with the 5.0L V-8 optional, and the Michelin TRX wheels/tires combination.

The most significant 1979 Mustang was the 1979½ Pace Car. As in 1964, Mustang was chosen for use in the 1979 Indy 500. Unlike in 1964, 11,000 pace car replicas were built. Engine availability was limited to the 302 or the Turbo four-cylinder. All pace cars had the same pewter/black paint treatment highlighted with orange and red tape stripes. Pace car lettering decals were dealer or customer installed. Other features were a rear spoiler, front air dam with integral foglamps, pop-up sunroof, rear facing non-functional hood scoop, TRX wheels and tires, and the interior sported Recaro seats finished in a unique pattern.

1979 Pace Car Ford Motor Co.

Chapter 16

1980 Mustang

Production Figures

66B 2dr Sedan	128,893	61R 3dr Hatchback Sport	98,497
66H 2dr Sedan Ghia	23,647	Total	271,322
61H 3dr Hatchback Ghia	20,285		

Serial Numbers
0F02A100001
0 — Last digit of model year
F — Assembly plant (F-Dearborn, R-San Jose)
02 — Plate code for Mustang (02-2dr sedan, 03-3dr hatchback, 04-2dr Ghia, 05-3dr Ghia)
A — Engine code
100001 — Consecutive unit number

Location
Stamped on riveted plate on driver's side of dash, visible through the windshield; certification label attached on rear face of driver's door.

Engine Codes
A — 2.3 liter 2V 4 cyl 88 hp(MT), 90 hp (AT)
W — 2.3 liter 2V 4 cyl (Turbocharged)
B — 3.3 liter 1V 6 cyl 91 hp(MT), 94 hp(AT)
D — 4.2 liter 2V V-8 119 hp

1980 Mustang Prices   Retail

2.3 liter 4 cyl Models	
2dr Sedan	$5,338.00
3dr Sedan	5,616.00
2dr Ghia	5,823.00
3dr Ghia	5,935.00
3.3 liter (200 cid) 6 cyl	256.00
4.2 liter (255 cid) 8 cyl, with Cobra Package	(144.00)
All others	338.00
2.3 liter 4 cyl Turbocharged	481.00
SelectShift automatic transmission	340.00
5-speed manual overdrive	156.00
Axle, optional ratio	18.00
Accent Group, exterior	63.00
Accent Group, interior, 2dr sedan	134.00
3dr sedan	120.00
Air conditioner, SelectAire	538.00
Battery, heavy-duty	20.00
Seatbelts, color-keyed deluxe	23.00

Bodyside protection, lower	34.00
Bracket, front license plate	N/C
Brakes, power front disc	78.00
Cargo area cover	44.00
Cobra Package	1,482.00
Cobra hood graphics	88.00
Console	166.00
Deflectors, mud & stone	25.00
Defroster, electric rear window	96.00
Emission system, Calif.	253.00
Emission system, high altitude	36.00
Exhaust system, sport-tuned	
With 2.3 liter turbocharged and automatic	N/C
With 4.2 liter engine	38.00
Glass, tinted complete	65.00
Heater, engine block immersion	15.00
Hood scoop	31.00
Light Group	41.00
Lock Group, power	113.00
Louvers, liftgate	141.00
Luggage carrier, roof-mounted	86.00
Mirror, lefthand remote-control	19.00
Mirrors, dual remote-control (black)	58.00
Moldings, rocker panel	30.00
Moldings, vinyl insert bodyside (narrow)	43.00
Moldings, vinyl insert bodyside (wide)	74.00
Paint, metallic glow	46.00
Paint treatment, lower Tu-tone	88.00
Premium sound system	94.00
Protection Group, appearance	
With front license plate bracket	41.00
Without bracket	38.00
Radio, AM delete	(93.00)
Radio, AM/FM monaural	53.00
Radio, AM/FM stereo	90.00
Radio, AM/FM stereo with cassette tape	179.00
Radio, AM/FM stereo with 8-track tape	166.00
Radio flexibility option	63.00
Roof, carriage	625.00
Roof, flip-up open air, Ghia or w/Light Group	204.00
All others	219.00
Roof, full vinyl	118.00
Seats, Recaro high-back buckets	531.00
Seat, 4-way manual driver's	38.00
Speakers, dual rear seat	38.00
Speed control, fingertip, 2dr without Sport option	129.00
All others	116.00
Sport option, with Carriage Roof	168.00
without Carriage Roof	186.00
Steering, power	160.00
Steering wheel, leather-wrapped	
2dr without Sport option	56.00

All others	44.00
Steering wheel, tilt	
2dr without Sport option	90.00
All others	78.00
Stripes, accent tape	
with Exterior Accent Group	19.00
without Accent Group	53.00
Stripes, pin (bodyside and deck lid)	34.00
Suspension, handling	35.00
Trim, accent cloth and vinyl seat	30.00
Trim, leather low-back bucket seat	349.00
Trim, cloth and vinyl bucket seat, Sedans	21.00
Ghia	46.00
Wheel covers, four turbine	
3dr or with Sport option	10.00
All others	43.00
Wheel covers, four wire	
3dr or with Sport option	89.00
Ghia	79.00
All others	121.00
Wheels, four cast aluminum	
3dr or with Sport option	289.00
Ghia	279.00
All others	321.00
Wheels, four forged metric aluminum	
3dr or with Sport option	323.00
Ghia models	313.00
All others	355.00
Wheels with trim rings, four styled steel	
3dr or with Sport option	71.00
Ghia	61.00
All others	104.00
Windshield wipers, interval	39.00
Wiper/washer, rear window	79.00
Limited Production options	
Floor mats, front color-keyed	19.00
Light, luggage compartment	5.00
Tinted glass, windshield	29.00
Tires, models having four 185/80Rx13 radial ply BSW as standard equipment, extra charge for four:	
P185/80Rx13 WSW	50.00
P175/75Rx14 BSW	25.00
P175/75Rx14 WSW	75.00
P185/75Rx14 BSW	50.00
P185/75Rx14 WSW	100.00
P185/75Rx14 RWL	117.00
TRX 190/65Rx390 BSW*	150.00
Models having four P175/75Rx14 Radial ply BSW as standard equipment, extra charge for four:	
P175/75Rx14 WSW	50.00
P185/75Rx14 BSW	25.00
P185/75Rx14 WSW	75.00

P185/75Rx14 RWL　　　　　　　　　　　　　　　92.00
TRX 190/65Rx390 BSW*　　　　　　　　　　　　125.00
*Require forged metric aluminum wheels

1980 Exterior Colors

Color	Code
Black	1C
Bright Blue	3J
Bright Yellow	6N
Polar White	9D
Silver Metallic	1G
Medium Grey Metallic	1P
Bright Caramel	5T
Dark Chamois Metallic	8A
Bright Bittersweet	2G
Light Medium Blue	3F
Dark Cordovan Metallic	8N
Bright Red	27
Bittersweet Glow	8D
Medium Blue Glow	3H
Chamois Glow	8W

Tu-tone Exterior Color Combinations

Upper color/lower color	Code
Dark Chamois Metallic/Chamois Glow	8A/8W
Polar White/Bright Yellow	9D/6N
Light Medium Blue/Bright Blue	3F/3J
Bright Bittersweet/Dark Cordovan Metallic	2G/8N
Polar White/Bittersweet Glow	9D/8D
Silver Metallic/Dark Cordovan Metallic	1G/8N
Chamois Glow/Dark Chamois Metallic	8W/8A
Silver Metallic/Medium Grey Metallic	1G/1P
Dark Cordovan Metallic/Bittersweet Glow	8N/8D
Bittersweet Glow/Dark Cordovan Metallic	8D/8N

1980 Interior Trim

Trim	Code
Black	A
Wedgewood Blue	B
Bittersweet	C
Red	D
White/Red	N
White/Blue	Q
Caramel	T
White/Black	W
Vaquero	Z
White/Caramel	2
White/Vaquero	9

1980 Mustang Facts

Very little was done in terms of styling in 1980. The biggest change was the downsizing of the 302 cI V-8 to an anemic 255 ci, putting out a paltry 119 hp. The cubic inch reduction was achieved by decreasing the bore from 4.00 inches to 3.68 inches. The 255 was only available with the automatic transmission.

Otherwise, engine choice remained the same—2.3L standard equipment with the 3.3L six and Turbocharged 2.3 four optional. You could also get the 2.3 Turbo with an automatic transmission.

The hot-dog Cobra model got the pace car's front and rear spoilers, rear opening (simulated) hood scoop, standard 2.3 Turbo engine (the 255 V-8 was optional), TRX wheels and tires, the sport-tuned exhaust system, dual black remote-control mirrors and the blacked-out treatment. The blacked-out treatment was extended to the interior as the dash panel got black engine-turned appliques.

The Recaro seats, first seen on the 1979½ Pace Car, were now optional on all Mustangs. They were a fairly expensive option at $531.00.

There were few changes with the luxury Ghia model, which got a restyled steering wheel. Emulating the convertible look was the Carriage Roof option available only on the two-door models.

The Sport option was still available on the three-door, consisting of styled steel wheels with trim rings, black rocker panel and window moldings, and bodyside moldings.

For the first time, the Mustang came equipped with halogen headlights, a maintenance-free battery and the P-metric radial tires.

New on the option list was the roof-mounted luggage carrier and cargo cover for the hatchbacks.

1980 three-door Cobra Ford Motor Co.

Chapter 17

1981 Mustang

Production Figures

66B 2dr Sedan	77,458	61H 3dr Hatchback	
66H 2dr Sedan Ghia	13,422	Ghia	14,273
61R 3dr Hatchback	77,399	Total	182,552

Serial Numbers

1FABP10A6BF000001
1FA — Ford Motor Co.
B — Restraint system (B-active belts)
P — Passenger car
10 — Body code (10/14-2dr sedan, 15-3dr, 12-2dr Ghia, 13-3dr Ghia)
A — Engine code
6 — Check digit which varies
B — Year (B-1981)
F — Plant (F-Dearborn)
000001 — Consecutive unit number

Location

Stamped on riveted plate on driver's side of dash, visible through the windshield; certification label attached on rear face of driver's door.

Engine Codes

A — 2.3 liter 2V 4 cyl 88 hp
B — 3.3 liter 1V 6 cyl 94 hp
D — 4.2 liter 2V V-8, 120 hp

1981 Mustang Prices

	Retail
2dr Sedan, P10/14H	$5,897.00
2dr Sedan, P10	6,363.00
3dr Sedan, P15	6,566.00
3dr Ghia, P12	6,786.00
3dr Ghia, P13	6,901.00
3.3L 6 cyl engine	213.00
4.2L 8 cyl engine	263.00
5-speed manual overdrive	183.00
SelectShift automatic	370.00
Traction-Lok axle	71.00
Tires, extra charge for four:	
P175/75Rx14 WSW	59.00
P185x75Rx14 BSW	28.00
P185x75Rx14 WSW	86.00
P185x75Rx14 RWL	107.00
190/65Rx390 TRX, requires forged metric wheels	135.00
Air conditioner, SelectAire	600.00

Battery, heavy-duty	22.00
Seatbelts, color-keyed deluxe	24.00
Bodyside protection lower	39.00
Bracket, front license plate	N/C
Brakes, power front disc	87.00
Cargo area cover	48.00
Cobra option	1,075.00
Cobra hood graphics	95.00
Cobra tape treatment, delete	(65.00)
Console	178.00
Deflectors, mud & stone	27.00
Defroster, rear window electric	115.00
Floor mats, front color-keyed	20.00
Glass, tinted complete	82.00
Hood scoop	35.00
Interior Accent Group, 2dr	159.00
3dr	145.00
Light Group	45.00
Lock Group, power	129.00
Louvers, liftgate	154.00
Luggage carrier, roof-mounted	99.00
Mirror, lefthand remote-control	22.00
Mirrors, dual remote-control, black	61.00
Moldings, rocker panel, black	30.00
Paint treatment, lower Tu-tone	96.00
Paint, Special Tu-tone	
Ghia	128.00
All others	165.00
Protection Group, appearance	44.00
Radio, AM/FM monaural	51.00
Radio, AM/FM stereo	88.00
Radio, AM with dual rear seat speakers	
(price for speakers)	39.00
Radio, AM/FM monaural with dual rear seat speakers	
(price for speakers)	39.00
Radio, AM/FM stereo with cassette tape	174.00
Radio, AM/FM stereo with 8-track tape	162.00
Radio flexibility option	65.00
Radio flexibility option with dual rear seat speakers	
(price for speakers)	39.00
Radio, AM delete	(61.00)
Sound system, premium	98.00
Roof, flip-up open air	
Ghia or with Light Group	227.00
All others	241.00
Roof, "T"	916.00
Speed control, fingertip	145.00
Sport option, with Carriage Roof	62.00
without Carriage Roof	82.00
Steering, power	176.00
Steering wheel, leather-wrapped	
2dr without Sport option	63.00

All others	51.00
Steering wheel, tilt	
2dr without Sport option	100.00
All others	88.00
Stripes, accent tape	57.00
Stripes, pin	37.00
Suspension, handling	46.00
Wheel covers, four turbine	46.00
Wheel covers, four wire	
Ghia	85.00
All others	132.00
Wheels, four cast aluminum	
Ghia	323.00
All others	370.00
Wheels, four forged metric aluminum	
Ghia	361.00
All others	407.00
Wheels with trim rings, four styled steel	
Ghia	67.00
All others	113.00
Windows, power side	152.00
Windshield wipers, interval	44.00
Wiper/washer, rear window	94.00
Emission system, Calif.	46.00
Emission system, high altitude	40.00
Exterior glow paint	50.00
Recaro high-back bucket seats	776.00
Accent cloth & vinyl seat trim	34.00
Cloth & vinyl bucket seat trim	
Sedans	23.00
Ghia	51.00
Leather low-back bucket seat trim	380.00
Full vinyl roof	127.00
Carriage roof	683.00
Limited Production options	
Floor mats, front carpet (color-keyed)	20.00
Glass, tinted, windshield only	29.00
Heater, engine block immersion	16.00
Light, luggage compartment	6.00

1981 Exterior Colors

Colors	Code	Colors	Code
Black	1C	Polar White	9D
Light Pewter Metallic	1T	Medium Pewter Metallic	17
Bright Bittersweet	2G	Red	24
Midnight Blue Metallic	3L	Bright Red	27
Dark Brown Metallic	5Q	Pastel Chamois	86
Bright Yellow	6N	Medium Blue Glow	3H
Dark Cordovan Metallic	8N	Bittersweet Glow	8D

Tu-tone Exterior Color Combinations

Upper color/lower color	Code
Light Pewter Metallic/Black	1T/1C
Medium Pewter Metallic/Black	17/1C
Bright Bittersweet/Black	2G/1C
Red/Black	24/1C
Bright Red/Black	27/1C
Bright Yellow/Black	6N/1C
Bittersweet Glow/Black	8D/1C
Dark Cordovan Metallic/Black	8N/1C
Pastel Chamois/Black	86/1C
Polar White/Black	9D/1C

Tu-tone Exterior Color Combinations

Upper color/lower color	Code
Medium Pewter Metallic/Light Pewter Metallic	17/1T
Red/Polar White	24/9D
Polar White/Bittersweet Glow	9D/8D

1981 Interior Trim

Trim	Code
Black	A
Wedgewood Blue	B
Red	D
Caramel	T
Pewter	P
Vaquero	Z

1981 Mustang Facts

Although early factory literature listed the 2.3L Turbo as an option, the engine was in fact dropped for reliability problems. Otherwise engine availability was the same as in 1980.

New options included a five-speed manual overdrive transmission (introduced during 1980), the Traction-Lok differential for the rear, rear window louvers for the hatchback and the T-roof, this time optional on both Mustang body styles.

Power windows were optional for the first time, too.

1981 was the last year of the Cobra option package, which continued basically unchanged from 1980.

1981 two-door Ford Motor Co.

Chapter 18

1982 Mustang

Production Figures

66B 2dr Sedan	45,316	61H 3dr Hatchback	
66H 2dr Sedan GLX	5,828	GLX	9,926
61B 3dr Hatchback	69,348	Total	130,418

Serial Numbers

1FABP10A6CF000001

1FA — Ford Motor Co.
B — Restraint system (B-active belts)
P — Passenger car
10 — Body code (10-2dr L/GL, 16-3dr GL/GT, 12-2dr GLX, 13-3dr GLX)
A — Engine code
6 — Check digit which varies
C — Year (C-1982)
F — Plant (F-Dearborn)
000001 — Consecutive unit number

Location

Stamped on riveted plate on driver's side of dash, visible through the windshield; certification label attached on rear face of driver's door.

Engine Codes

A — 2.3 liter 2V 4 cyl 88 hp
B — 3.3 liter 1V 6 cyl 94 hp
D — 4.2 liter 2V V-8 120 hp
F — 5.0 liter 2V V-8 157 hp

1982 Mustang Prices	Retail
2dr L Sedan, P10	$6,345.00
2dr GL Sedan, P10	6,844.00
3dr GL Sedan, P16	6,979.00
2dr GLX Sedan, P12	6,980.00
3dr GLX Sedan, P13	7,101.00
3dr GT Sedan, P16	8,397.00
3.3L 6 cyl engine	213.00
4.2L 8 cyl engine	
GL	(57.00)
All others	283.00
5.0L 8 cyl engine (std. on GT)	
Use with TR Performance Package	494.00
All others	544.00
5-speed manual overdrive	196.00
SelectShift automatic	411.00
Optional ratio	N/C

Traction-Lok axle	76.00
Tires, extra charge for four (except GT):	
P175/75Rx14 WSW	72.00
P185/75Rx14 BSW	36.00
P185/75Rx14 WSW	108.00
P185/75Rx14 RWL	128.00
GT, extra charge for four:	
P185/75Rx14 WSW	72.00
P185/75Rx14 RWL	91.00
Air conditioner, SelectAire	688.00
Battery, heavy-duty	24.00
Bodyside protection, lower	41.00
Bracket, front license plate	N/C
Brakes, power front disc	93.00
Cargo area cover	51.00
Console	191.00
Defroster, rear window electric	124.00
Glass, tinted complete	88.00
Hood scoop	38.00
Light Group	49.00
Lock Group, power	139.00
Louvers, liftgate	165.00
Mirror, righthand remote-control	41.00
Moldings, rocker panel, black	33.00
Paint treatment, lower Tu-tone	116.00
Paint, Special Tu-tone	
GL & GLX	150.00
L	189.00
Protection Group, appearance	48.00
Radio, AM/FM monaural	76.00
Radio, AM with dual rear seat speakers	
(price for speakers)	39.00
Radio, AM/FM monaural with dual rear seat speakers	
(price for speakers)	39.00
Radio, AM/FM stereo with cassette tape	184.00
Radio, AM/FM stereo with 8-track tape	184.00
AM radio credit option	(61.00)
Sound system, premium	117.00
Roof, flip-up open air	276.00
Roof, "T"	1,021.00
Speed control, fingertip	155.00
Steering, power	202.00
Steering wheel, leather-wrapped	55.00
Steering wheel, tilt	95.00
Stripes, accent tape	62.00
Suspension, handling	50.00
TR Performance Suspension Package	
L	589.00
GL & GLX	539.00
GT	111.00
Wheel covers, four wire	
L	148.00

GL & GLX	98.00
Wheels, four cast aluminum	
L	404.00
GL & GLX	354.00
Wheels with trim rings, four styled steel	
L	128.00
GL & GLX	78.00
Windows, power side	165.00
Windshield wipers, interval	48.00
Wiper/washer, rear window	101.00
Emission system, Calif.	46.00
Emission system, high altitude	N/C
Metallic glow paint	54.00
Recaro high-back bucket seats	834.00
Cloth & vinyl seat trim	
L	29.00
GL	40.00
GLX & GT	57.00
Leather, low bucket seat trim	415.00
Full vinyl roof	149.00
Carriage roof	746.00
Limited Production options	
Floor mats, front carpet, color-keyed	22.00
Glass, tinted, windshield only	32.00
Heater, engine block, immersion	17.00
Light, luggage compartment	7.00

1982 Exterior Colors

Color	Code
Black	1C
Silver Metallic	1G
Medium Grey Metallic	1P
Red	24
Bright Red	27
Dark Blue Metallic	3D
Medium Vanilla	6Y
Pastel Vanilla	6Z
Medium Yellow	61
Dark Curry Brown Metallic	69
Dark Cordovan Metallic	8N
Polar White	9D
Medium Blue Glow	3H
Bittersweet Glow	8D

Tu-tone Exterior Color Combinations

Upper color/lower color	Code
Medium Grey Metallic/Silver Metallic	1P/1G
Dark Blue Metallic/Medium Blue Glow	3D/3H
Medium Vanilla/Pastel Vanilla	6Y/6Z
Dark Cordovan Metallic/Bittersweet Glow	8N/8D
Silver Metallic/Black	1G/1C
Medium Grey Metallic/Black	1P/1C
Red/Black	24/1C
Bright Red/Black	27/1C
Dark Blue Metallic/Black	3D/1C
Medium Blue Glow/Black	3H/1C
Medium Vanilla/Black	6Y/1C
Pastel Vanilla/Black	6Z/1C
Medium Yellow/Black	61/1C
Dark Cordovan Metallic/Black	8N/1C

Tu-tone Exterior Color Combinations

Upper color/lower color	Code
Bittersweet Glow/Black	8D/1C
Polar White/Black	9D/1C

1982 Interior Trim

Trim	Code
Black	A
Wedgewood Blue	B
Red	D
Vaquero	Z
White/Vaquero	9
Opal/Red	N
French Vanilla	V
Opal/Black	W
Opal/Vaquero	9

1982 Mustang Facts

The most significant change of 1982 was the reintroduction of the Mustang GT and the 302 ci V-8, dubbed the 5.0L H.O. Although the 5.0L engine could be had on any Mustang, the Mustang GT model included the engine as standard equipment along with a host of other options.

The resurrected 5.0L came with stouter internal components such as a double roller timing chain, and a higher lift cam (as compared to the 1979 version).

Mandatory options with the 5.0 included the four-speed manual overdrive, 3.08 Traction-Lok rear, power brakes, power steering and the handling suspension which included traction bars for the 5.0. Fourteen-inch cast aluminum wheels with P185/75R14 tires were standard; the TRX wheels/tires/suspension was optional.

The GT also came with a slightly redesigned grille, forward facing non-functional hood scoop and the pace car air dam and rear spoiler. Color choice was limited to three—Red, Black or Metallic Silver with red or black interiors. Extensive use of black paint in the interior and blacked-out treatment on the exterior served to enhance the Mustang's new-found performance image.

All Ford engines beginning in 1982 were painted a light gray.

1982 Mustang GT Hatchback Ford Motor Co.

Chapter 19

1983 Mustang

Production Figures

66B 2dr Sedan	33,201	61B 3dr Hatchback	64,234
66B 2dr Convertible	23,438	Total	120,873

Serial Numbers

1FABP26A6DF000001
1FA — Ford Motor Co.
B — Restraint system (B-active belts)
P — Passenger car
26 — Body code (26-2dr sedan, 27-2dr convertible, 28-3dr hatchback)
A — Engine code
6 — Check digit which varies
D — Year (D-1983)
F — Assembly plant (F-Dearborn)
000001 — Consecutive unit number

Location
Stamped on plate riveted on driver's side of dash, visible through the windshield; certification label attached on rear face of driver's door.

Engine Codes
A — 2.3L 1V 4 cyl 88 hp
T — 2.3L EFI 4 cyl 145 hp (Turbo GT)
3 — 3.8L 2V V-6 112 hp
M — 5.0L 4V V-8 175 hp (HO)

1983 Mustang Prices

	Retail
2dr L Sedan, P26	$6,727.00
2dr GL Sedan, P26/60C	7,264.00
3dr GL Sedan, P28/60C	7,439.00
2dr GLX Sedan, P26/602	7,398.00
2dr GLX Convertible, P27/602	12,467.00
3dr GLX Sedan, P28/602	7,557.00
3dr GT Sedan, P28/932	9,449.00
3dr Turbo GT, P28/932	9,714.00
2dr GT Convertible, P27/932	13,479.00
3.8L 6 cyl (std. GLX Convertible)	309.00
5.0L 4V HO 8 cyl package, 5-speed transmission (std. on GT, except Turbo GT)	
GLX Convertible	719.00
All others	1,467.00
5-speed manual overdrive (std. GT & 5.0L)	124.00
SelectShift automatic (std. GLX Convertible)	439.00

4-speed manual credit (for GT or with 5.0L, NA Turbo GT)	(124.00)
Optional axle ratio	N/C
Traction-Lok axle	95.00
Tires, extra charge for four (except GT & models with 5.0L):	
P185/75Rx14 WSW	72.00
P195/75Rx14 WSW	108.00
P205/70HRx14 BSW	224.00
Michelin TRX P220/55R 390 BSW	551.00
GT (credit)	(27.00)
Models with 5.0L	327.00
Air conditioner, SelectAire	737.00
Battery, heavy-duty	26.00
Bodyside protection, lower	41.00
Bracket, front license plate	N/C
Brakes, power front disc	93.00
Console	191.00
Defroster, rear window electric	135.00
Glass, tinted complete	105.00
Light Group	55.00
Lock Group, power	172.00
Louvers, liftgate	171.00
Mirror, righthand remote-control	44.00
Moldings, rocker panel, black	39.00
Paint treatment, lower Tu-tone	116.00
Paint, special Tu-tone	
GL & GLX	150.00
L	189.00
Protection Group, appearance	39.00
Radio, AM/FM monaural	82.00
Radio, AM/FM stereo	109.00
Radio, AM/FM stereo with cassette tape	199.00
Radio, AM/FM stereo with 8-track tape	199.00
AM radio credit option	(61.00)
Sound system, premium	117.00
Roof, flip-up open air	310.00
Roof, "T"	1,074.00
Speed control, fingertip	170.00
Steering, power	202.00
Steering wheel, leather-wrapped	59.00
Steering wheel, tilt	105.00
Suspension package, handling	252.00
Wheel covers, four turbine	N/C
Wheel covers, four wire	
L	148.00
GL & GLX	98.00
Wheels, four cast aluminum	
L	404.00
GL & GLX	354.00
Wheels with trim rings, four styled steel	
L	128.00

GL & GLX	78.00
Windows, power front	193.00
Windshield wipers, interval	49.00
Delete standard accent stripe	N/C
Emissions system, Calif.	76.00
Emissions system, high altitude	N/C
Metallic glow paint	54.00
Cloth sports performance seats	196.00
Cloth & vinyl seat trim	
L	29.00
GL	40.00
GLX and GT	57.00
Leather low-back bucket seat trim	415.00
Limited Production options	
Floor mats, front carpet, color-keyed	22.00
Glass, tinted, windshield only	38.00
Heater, engine block immersion	17.00

1983 Exterior Colors

Colors	Code
Black	1C
Polar White	9D
Silver Metallic	1G
Medium Charcoal Metallic	1B
Red	24
Bright Red	27
Dark Academy Blue Metallic	3D
Dark Walnut Metallic	5U
Light Desert Tan	8Q
Medium Yellow	61
Midnight Blue Metallic	3L
Light Academy Blue Glow	38
Desert Tan Glow	9N
Bright Bittersweet	2G

1983 Interior Trim

Trim	Code
Black	A
Cadet Blue	B
Red	D
Walnut	E
Opal/Red	N
Opal/Blue	Q
Opal/Black	W
Opal/Walnut	Z

1983 Mustang Facts

The 1983 Mustang featured a new front end treatment and new rear taillight lenses. A third body style, the convertible in GL or GT forms, was added. Mechanically, the Mustang was the focus of considerable improvement. In spite of all this, the 1983 Mustang had the distinction of being the lowest-selling up to that time.

Model configuration was similar to 1982. The base Mustang was the L two-door. The upgraded GL was available in two-door, three-door and convertible while the GLX could be had in two- or three-door versions. The 3dr GT was joined by the three-door Turbo GT and the two-door GT convertible.

The 2.3L four-cylinder was the base engine. The 3.3L inline was finally replaced by a new V-6, the 3.8L. This was an all-new design which featured aluminum cylinder heads. Porting was similar to that found on the 351C 2V heads, though the valves were not canted. Rated at 112 hp, it came only with the SelectShift automatic transmission.

The 2.3L Turbo surfaced again, this time as part of the Turbo GT, meaning that you could only get this engine if you ordered the three-door Turbo GT. It was not available with any other Mustang model. The 2.3L Turbo GT, rated at 145 hp, was superior to Ford's earlier turbocharging attempt. The engine featured Electronic Fuel Injection and its earlier reliability problems had been licked. However, it was overshadowed by the born-again 5.0L HO.

Development continued on the 5.0L HO. In 1983 guise, the two-barrel carburetor was trashed in favor of a Holley 600 cfm four-barrel mounted on an aluminum intake manifold—just like the old days! Horsepower jumped to 175.

To improve the GT's handling, the rear stabilizer bar was increased slightly in size, to .67 inch. However, the 1983's improvement in handling mostly can be attributed to larger sized tires. The standard GT tires went from a 185/75Rx14 to a 205/70HTx14 which was lower in profile and wider. Similarly, the TRX Michelins were increased to a P220/55R 390.

The GT's hood scoop changed ends again; this time the opening faced the rear.

The convertible, the first in ten years, came with a power top that had a glass backlite. Standard was the 3.8L V-6 and all the GLX features. Also available was the GT convertible. Convertibles were built by Cars & Concepts for Ford.

1983 was the last year that the 8-track player was offered, due to the waning popularity of the 8-track tape medium.

Two roof treatments were available—a flip-up open air sunroof and the T-tops. The Carriage Roof option, no longer necessary, was deleted from the option list.

1983 convertible GT Ford Motor Co.

Chapter 20

1984 Mustang

Production Figures

66B 2dr Sedan	37,680
66B 2dr Convertible	17,600
61B 3dr Hatchback	86,200
Total	141,480

1984 Mustang 20th Anniversary Edition

3dr Turbo GT	350
3dr 5.0L GT	3,333
2dr Convertible Turbo GT	104
2dr Convertible 5.0L GT	1,213
Ford VIP Convertibles	15
Ford of Canada (total)	245
Total	5,260

Source: 20th Anniversary Registry

Serial Numbers

1FABP26A6DF000001

1FA — Ford Motor Co.
B — Restraint system (B-active belts)
P — Passenger car
26 — Body code (26-2dr sedan, 27-2dr convertible, 28-3dr hatchback)
A — Engine code
6 — Check digit which varies
D — Year (D-1984)
F — Assembly plant (F-Dearborn)
000001 — Consecutive unit number

Location
Stamped on riveted plate on driver's side of dash, visible through the windshield; certification label attached to rear face of driver's door.

Engine Codes
A — 2.3L 1V 4 cyl 88 hp
T — 2.3L EFI 4 cyl 145 hp (Turbo GT)
W — 2.3L EFI 4 cyl 175 hp (SVO)
3 — 3.8L EFI V-6 120 hp
F — 5.0L EFI V-8 165 hp
M — 5.0L 4V V-8 175 hp (HO)

1984 Mustang Prices

	Retail
2dr L Sedan, P26	$7,089.00
3dr L Sedan, P28	7,260.00
2dr LX Sedan, P26/602	7,281.00
3dr LX Sedan, P28/602	7,487.00
2dr LX Convertible, P27/602	11,840.00
3dr GT Sedan, P28/932	9,774.00
2dr GT Convertible, P27/932	13,247.00
3dr Turbo GT Sedan, P28/932	9,958.00

2dr Turbo GT Convertible, P27/932	13,441.00
3dr SV0 P28/939/99T	15,585.00
3.8L EFI 6 cyl engine (std. LX Convertible)	409.00
5.0L 4V H0 8 cyl package/5-speed transmission (std. on GT except Turbo GT)	
LX Convertible	727.00
All other models	1,574.00
SelectShift automatic transmission (std. LX Convertible)	439.00
Automatic overdrive	551.00
Optional axle ratio	N/C
Traction-Lok axle	95.00
Tires, extra charge for four:	
P185/75Rx14 WSW	72.00
P195/75Rx14 WSW	109.00
P205/70VRx14 BSW	224.00
Air conditioner, SelectAire	743.00
Battery, heavy-duty	27.00
Bodyside protection, lower	41.00
Bracket, front license plate	N/C
Brakes, power front disc	93.00
Competition Preparation option (SV0)	(1,253.00)
Console	191.00
Defroster, rear window electric	140.00
Floor mats, front carpet color-keyed	22.00
Glass, tinted complete	110.00
Heater, engine block immersion	18.00
Light Convenience Group	
LX and GT	55.00
All other models	88.00
Lock Group, power	177.00
Mirror, righthand remote-control	46.00
Moldings, rocker panel, black	39.00
Paint treatment, lower Tu-tone	116.00
Paint, special Tu-tone	
L	189.00
LX	150.00
Radio, AM/FM stereo	109.00
Radio, AM/FM stereo with cassette tape	222.00
AM radio, credit option	(39.00)
Sound system, premium	151.00
Roof, flip-up open air	315.00
Roof, "T"	1,074.00
Speed control, fingertip	176.00
Steering, power	202.00
Steering wheel, tilt	110.00
Suspension package, handling	252.00
Wheel covers, four wire style	98.00
Wheels, four cast aluminum	354.00
Wheels, four cast metric aluminum, P220/55R390 Michelin TRX BSW tires	
L & LX	551.00

With 5.0L engine	327.00
GT	(27.00)
Wheels with trim rings, four styled steel	78.00
Windows, power side	
Convertible	272.00
All others	198.00
Windshield wipers, interval	50.00
Emission system, Calif.	99.00
Emission system, high altitude	N/C
Metallic glow paint	54.00
Cloth articulated sport seats	366.00
Vinyl seat trim	
L	29.00
LX & GT	29.00
Leather seat trim (Convertible only)	415.00
SVO	189.00

1984 Exterior Colors	Code	1984 Exterior Colors	Code
Black	1C	Desert Tan Glow	9J
Silver Metallic	1E		
Bright Canyon Red	27	**1984 Interior Trim**	**Code**
Dark Academy Blue Metallic	5C	Charcoal	A
Light Desert Tan	8Q	Cadet Blue	B
Oxford White	9L	Canyon Red	D
Dark Charcoal Metallic	9W	Desert Tan	H
Medium Canyon Red Glow	2B	White/Red (Convertible)	N
Light Academy Blue Glow	35	White/Blue (Convertible)	Q
Bright Copper Glow	9C	White/Charcoal (Convertible)	W

1984 Mustang Facts

Model lineup changed for 1984. The base two-door and three-door Mustangs were L models. Upgraded two-door and three-door models, including the convertible, were LX Mustangs. GT Mustangs were available in three-door and convertible form, as were Turbo GT Mustangs. Midway through the model year, the 1984½ SVO Mustang was introduced. SVO stands for Special Vehicle Operations, a Ford unit formed in 1981 which was assigned the task of building a performance parts program and developing special high-performance street cars.

Exterior and interior styling was basically a carryover from 1983. Engine availability was the same, too, but there were some changes on the 3.8L V-6 and 5.0L HO. The 3.8L V-6 came with throttle body Electronic Fuel Injection and, again, this was the standard engine on the LX convertible. GT Mustangs with the automatic transmission got Ford's new four-speed automatic overdrive but the engine was equipped with a new intake system. Rather than using the 600 cfm Holley carburetor, it came with throttle body EFI, which brought its horsepower rating down to

165. Factory literature listed dual exhaust versions of the 5.0L as a mid-year introduction, but these were never made available.

The Turbo GT Mustangs were unchanged from 1983.

The standard transmission on the GTs was the Borg-Warner T-5, a five-speed manual overdrive, which gave the Mustang GT (and SVO) driver far more flexibility.

The big news was the 1984½ SVO Mustang, designed to deliver superior performance while appealing to a more sophisticated buyer. The heart of the SVO was yet another version of the 2.3L four-cylinder. Similar to the Turbo GT engine, this engine came with an intercooler and EFI to boost hp to 175. Maximum boost was 14 psi, which was electronically controlled. Other features included a revised front suspension with Koni adjustable shocks, four wheel disc brakes, the Quadra-Shock rear suspension (from the Thunderbird Turbo Coupe) and 16x7 inch aluminum wheels with P225/50VR Goodyear NCT tires. The SVO was equipped with articulated front seats with adjustable lumbar support, leather-wrapped steering wheel and a premium stereo system. Major options were air conditioning, power windows and locks, cassette, pop-up sunroof and leather interior. Exterior colors were black, charcoal, silver and red, all with a charcoal interior. Externally, the SVO got a unique grille, functional hood scoop, wheel spats in front of the rear wheelwell openings and a large biplane rear spoiler. It was a nice package, but its high price—some $6,000 more than the Mustang GT—limited its appeal.

The other special of 1984 was the 20th Anniversary Edition Mustang which commemorated the Mustang's 20th anniversary. They were all Mustang GTs, three-doors and convertibles powered by the 302 HO or Turbo GT. Paint color was Oxford White with Canyon Red interiors. The 20th also came with articulated front seats but without the adjustable lumbar support. The rest of the package consisted of a rocker panel tape treatment with GT350 lettering, original (1965) type front fender emblems and two 20th Anniversary dash panel badges. The first of these, a horseshoe medallion, was located on the passenger's side of the dash. About three to four months after purchase, the owner was sent a form to fill out so that the second medallion could be obtained. This one read "Limited Edition" followed by a serial number (unrelated to the car's VIN) and, below that, the owner's name. A total of 5,260 Anniversary Editions were built.

The standard GT tires were upgraded to a V rating, good for 130 mph and above. The TRX wheels/tires were still optional, but it was their last year. From 1985 on, Goodyear Eagles were used on the GT.

The adjustable articulated seats were optional on all Mustangs.

Ford took over production of the convertible Mustangs from Cars & Concepts, making 1984 and later convertibles true factory convertibles.

From 1984 on, Mustangs were equipped with Ford's EEC-IV (electronic engine control system), monitoring all engine functions. The EEC-IV was designed to meet emissions regulations while maximizing performance.

1984 20th Anniversary convertible

1984 20th Anniversary three-door Stephen Jacobs

1984½ SVO Ford Motor Co.

Chapter 21

1985 Mustang

Production Figures

66B 2dr Sedan	56,781	61B 3dr Hatchback	84,623
66B 2dr Convertible	15,110	Total	156,514

Serial Numbers

1FABP26A6FF000001

1FA — Ford Motor Co.
B — Restraint system (B-active belts)
P — Passenger car
26 — Body code (26-2dr sedan, 27-2dr convertible, 28-3dr hatchback)
A — Engine code
6 — Check digit which varies
F — Year (F-1985)
F — Assembly plant (F-Dearborn)
000001 — Consecutive unit number

Location
Stamped on riveted plate on driver's side of dash, visible through the windshield; certification label attached on rear face of driver's door.

Engine Codes
A — 2.3L 1V 4 cyl 88 hp
W — 2.3L EFI 4 cyl 205 hp (SVO)
3 — 3.8L EFI V-6 120 hp
M — 5.0L EFI/4V V-8 165/210 hp (HO)

1985 Mustang Prices

	Retail
2dr LX Sedan, P26/602	$6,989.00
3dr LX Sedan, P28/602	7,509.00
2dr LX Convertible, P27/602	12,237.00
3dr GT Sedan, P28/932	10,224.00
2dr GT Convertible, P27/932	13,930.00
3dr SVO, P28/937	14,806.00
3.8L EFI 6 cyl engine (std. LX Convertible)	454.00
5.0L 4V HO 8 cyl package (std. on GT)	
LX Convertible	172.00
LX Sedans	1,020.00
5-speed manual overdrive (std. GT & SVO)	124.00
SelectShift automatic transmission (std. LX Convertible)	470.00
Automatic overdrive	
LX	706.00
GT	582.00
Optional axle ratio	N/C

Traction-Lok axle	100.00
Tires, extra charge for four:	
P205/70Rx14 WSW	109.00
P205/70VRx14 BSW performance	238.00
P225/60VRx15 BSW cast aluminum wheels	665.00
Air conditioner, manual	762.00
Battery, heavy-duty	27.00
Bracket, front license plate	N/C
Competition Preparation option (SVO)	(1,451.00)
Console	191.00
Defroster, rear window electric	145.00
Glass, tinted complete	115.00
Heater, engine block immersion	18.00
Light Group	55.00
Lock Group, power	
LX	215.00
GT	182.00
Paint treatment, lower Tu-tone	116.00
Radio, electronic AM/FM stereo with cassette tape	300.00
Radio, credit option	(148.00)
Sound system, premium	138.00
Roof, flip-up open air	315.00
Roof, "T"	1,100.00
Speed control, fingertip	176.00
Spoiler, single wing	N/C
Steering wheel, tilt	115.00
Wheel covers, four wire style	98.00
Wheels, four styled road	178.00
Windows, power side	
Convertible	282.00
All others	207.00
Emissions system, Calif.	99.00
Emission system, high altitude	N/C
Leather articulated sport seats	
LX Convertible	780.00
GT Convertible	415.00
Vinyl seat trim	29.00
Leather seat trim, SVO	189.00

1985 Exterior Colors

Color	Code
Black	1C
Medium Charcoal	1B
Canyon Red	2C
Pastel Regatta Blue	3M
Oxford Grey	1U
Sand Beige	8L
Silver	1Q
Jalapena Red	2R
Oxford White	9L
Dark Sable	8Y
Medium Regatta Blue	3Y

1985 Interior Trim

Trim	Code
Charcoal	A
Regatta Blue	B
Canyon Red	D
Sand Beige	Y
White/Red (Convertible)	N
White/Blue (Convertible)	Q
White/Charcoal (Convertible)	W

1985 Mustang Facts

Emulating the SVO's look, all 1985 Mustangs got a new frontal treatment along with different bodyside moldings. Interior configuration was still the same. The Turbo GT was dropped, but the SVO was still available. The SVO was updated during the model year and rereleased as a 1985½ model.

Besides the SVO, Mustang models were limited to two-door, three-door and convertible in LX trim with the GT available as a three-door or convertible. Of course, the LX could be optioned out with the 5.0L HO and the GT's 15x7 wheels/tires to GT specs but at a higher cost.

Engine choice was limited to the standard 2.3L, the optional 3.8L V-6 (standard on the LX convertible) and the powerhouse 5.0L HO.

The biggest and most significant changes occurred on the Mustang GT as the muscle-car wars, 1980's style, were heating up.

The engine compartment of the GT looked outwardly the same. Upon a closer look the new exhaust headers were evident. These were stainless steel headers and not the typical so-called free-flowing cast iron exhaust manifolds. Coupled to a true dual exhaust system, with dual catalytic converters, horsepower jumped to 210. Internally, the 5.0L got a hotter camshaft with roller lifters. Contributing to the 210 hp total was a new accessory drive system which slowed down the air conditioner, alternator and power steering pump to half speed above idle.

The T-5 transmission got shorter gear throws for quicker response but what really improved the GT's road manners was the addition of the Quadra-Shock rear, a larger rear stabilizer bar and the new Goodyear tires. These were the Goodyear Eagle P225/60VR-15 unidirectional Gatorbacks mounted on new 15x7 inch aluminum wheels. These big tires, wider than any previous Mustang tires, helped the Mustang handle better.

In the interior, the articulated seats were standard as was the SVO's steering wheel. The exterior GT graphics were slightly subdued, using a dark charcoal treatment rather than the previous blacked-out treatment.

The GT was also available with an automatic overdrive transmission; however it did not use the same engine. This was the EFI version of the 5.0L rated at 165 hp. This engine used throttle body injection which is not to be confused with the multi-port injection available from 1986 on.

The SVO was updated as a mid-year model. Visually, the SVO got flush headlamps, but the important changes were not visual. The engine, through the use of a higher performance camshaft, reworked intake manifold and Turbocharger, a freer flowing exhaust system, larger fuel injection nozzles and a one-pound increase in Turbo boost, was rated at an impressive 205 hp. Through the use of redesigned brackets, the engine was far smoother, too. The suspension was tighter with 14.7:1 ratio steering, stiffer shocks and teflon-lined stabilizer bar bushings which contributed to improved handling, as did the Goodyear Eagle tires replacing the NCTs.

The rear wing spoiler (GT-type) was a no-cost option on the LX three-door hatchbacks.

The T-roofs were still available as was the flip-up open air sunroof.

Ninety Twister IIs were sold through the Kansas City sales district. Strictly a cosmetic package, it consisted of exterior graphics and a special dash plaque. Seventy-six were three-door GTs while fourteen were GT convertibles. Colors were limited to Bright Red, Medium Canyon Red, Oxford White and Silver Metallic.

1985–86 three-door

Chapter 22

1986 Mustang

Production Figures

66B 2dr Sedan	83,774	61B 3dr Hatchback	117,690
66B 2dr Convertible	22,946	Total	224,410

Serial Numbers
1FABP26A6GF000001
1FA — Ford Motor Co.
B — Restraint system (B-active belts)
P — Passenger car
26 — Body code (26-2dr sedan, 27-2dr convertible, 28-3dr hatchback)
A — Engine code
6 — Check digit which varies
G — Year (G-1986)
F — Plant (F-Dearborn)
000001 — Consecutive unit number

Location
Stamped on riveted plate on driver's side of dash, visible through the windshield; certification label attached on rear face of driver's door.

Engine Codes
S — 2.3L 1V 4 cyl 88 hp
W — 2.3L EFI 4 cyl 205 hp (SVO)
3 — 3.8L EFI V-6 120 hp
M — 5.0L EFI V-8 200 hp (HO)

1986 Mustang Prices

	Retail
2dr LX Sedan, P26	$7,420.00
2dr LX Hatchback, P28	7,974.00
2dr LX Convertible, P27	13,214.00
2dr GT Hatchback, P28	11,102.00
2dr GT Convertible, P27	14,945.00
2dr SVO, P28	15,272.00
3.8L EFI 6 cyl (std. LX Convertible)	565.00
5.0L EFI HO 8 cyl package (std. GT)	
LX Convertible	646.00
LX Sedan or Convertible	1,211.00
5-speed manual overdrive (std. GT)	
LX Sedan or Hatchback	124.00
LX Convertible	(410.00)
SelectShift automatic (std. LX Convertible)	534.00

Automatic overdrive	
LX Sedan or Hatchback	771.00
LX Convertible	237.00
GT	646.00
Tires, extra charge for four:	
P205/70Rx14 WSW	118.00
P225/60VRx15 BSW cast aluminum wheels	674.00
Air conditioner, manual	788.00
Battery, heavy-duty	27.00
Bracket, front license plate	N/C
Console	191.00
Competition Preparation option (SVO)	1,451.00
Defroster, rear window	145.00
Glass, tinted complete	120.00
Heater, engine block immersion	18.00
Hood graphic credit	N/C
Light Group	55.00
Lock Group, power	
LX	244.00
GT	206.00
Paint, lower charcoal accent	116.00
Radio, AM/FM stereo with cassette tape	157.00
Radio, electronic AM/FM stereo with cassette tape	310.00
Radio, credit option	(157.00)
Sound system, premium	138.00
Roof, flip-up open air	355.00
Roof, "T"	1,120.00
Speed control	176.00
Spoiler, single rear wing	N/C
Steering wheel, tilt	124.00
Wheel covers, four wire style	98.00
Wheels, four styled road	178.00
Windows, power side	
Convertibles	296.00
All others	222.00
Emission system, Calif.	99.00
Emission system, high altitude	N/C
Leather articulated sport seats	
LX Convertible	780.00
GT Convertible	415.00
Vinyl seat trim	29.00

1986 Exterior Colors

Colors	Code
Black	1C
Dark Grey Metallic	1B
Silver Metallic	1E
Medium Canyon Red Metallic	2C
Bright Red	2A
Light Regatta Blue Metallic	3J
Dark Sage	4E
Dark Slate Metallic	4M
Shadow Blue Metallic	7B
Sand Beige	8A
Dark Clove Metallic	8Y
Oxford White	9L

1986 Interior Trim	Code
Charcoal	A
Regatta Blue	B
Canyon Red	D
Sand Beige	Y
White/Red (Convertible)	N
White/Blue (Convertible)	Q
White/Charcoal (Convertible)	W

1986 Mustang Facts

1986 was the last year for the SVO. For the typical Mustang buyer, the 302 HO powered GT provided similar performance for a lot less money.

Model line-up was simplified—LX Mustangs could be had in a two-door, three-door and convertible, and the GT in either the three-door hatchback or convertible. This simplification process would be continued in the following years with the number of options available decreasing each year.

Besides minor trim and color changes, Mustang styling was unchanged from 1985.

The 5.0L HO got a new intake setup—sequential multi-port electronic fuel injection—but a lower horsepower rating than 1985, 200 hp versus 210 hp. Other engine changes included redesigned cylinder heads, beefed up block and a more efficient water pump.

The GT's hood graphics could now be deleted—a no-charge option. The rear wing spoiler for the hatchbacks was still a no-cost option.

The 7.5 inch rear was replaced by the stronger 8.8 in. integral carrier unit from Ford's full size cars on the GT. Standard ratio was 2.73:1 with 3.08:1 optional. The 3.27:1 ratio was optional only with the automatic overdrive transmission.

Chapter 23

1987 Mustang

Production Figures
66B 2dr Sedan	43,257	61B 3dr Hatchback	**94,441**
66B 2dr Convertible	32,074	Total	**159,145**

Serial Numbers
1FABP40A6HF000001
1FA — Ford Motor Co.
B — Restraint system (B-active belts)
P — Passenger car
40 — Body code (40-2dr LX, 41-3dr LX, 42-3dr GT, 44-2dr LX convertible, 45-2dr GT convertible)
A — Engine code
6 — Check digit which varies
H — Year (H-1987)
F — Plant (F-Dearborn)
000001 — Consecutive unit number

Location
Stamped on riveted plate on driver's side of dash, visible through the windshield; certification label attached to left B-pillar.

Engine Codes
A — 2.3L 1V 4 cyl 88 hp
M — 5.0L EFI V-8 225 hp (HO)

1987 Mustang Prices

	Retail
2 dr LX Sedan, P40	$8,271.00
2 dr LX Hatchback, P41	8,690.00
2 dr LX Convertible, P44	13,052.00
2 dr GT Hatchback, P42	12,106.00
2 dr GT Convertible, P45	15,852.00
5.0L EFI HO 8 cyl package (std. GT)	1,885.00
Automatic overdrive	
LX	515.00
GT	515.00
Tires, extra charge for four P195/75Rx14 WSW	82.00
Air conditioner, manual	788.00
Battery, heavy-duty	27.00
Bracket, front license plate	N/C
Defroster, rear window	145.00
Heater, engine block immersion	18.00
Lock Group, power	
LX	244.00
GT	206.00
Mirrors, dual electric remote	60.00

Mirrors, dual illuminated visor	100.00
Molding, bodyside insert stripe	49.00
Radio, electronic AM/FM with cassette tape	137.00
Sound system, premium	168.00
Radio credit option	(206.00)
Graphic equalizer	218.00
Roof, flip-up open air	355.00
Roof, "T", LX	1,798.00
GT	1,618.00
Speed control	176.00
Steering wheel, tilt	124.00
Wheel covers, four wire style	98.00
Wheels, four styled road	178.00
Windows, power side	
Sedan & Hatchback	222.00
Convertibles	296.00
Emission system, Calif.	99.00
Emission system, high altitude	N/C
Leather articulated sport seats	
LX Convertible	780.00
GT Convertible	415.00
Vinyl seat trim	29.00
Lower titanium accent treatment (GT only)	N/C

1987 Exterior Colors

Color	Code
Black	1C
Dark Grey Metallic	9R
Light Grey	1K
Scarlet Red	2D
Medium Cabernet	2H
Medium Shadow Blue Metallic	3R
Medium Yellow	66
Dark Shadow Blue Metallic	7N
Bright Regatta Blue Metallic	7H
Sand Beige	8L
Dark Clove Metallic	5H
Oxford White	9L

1987 Interior Trim

Trim	Code
Regatta Blue	B
Scarlet	D
Smoke	G
Sand Beige	Y
White/Scarlet (Convertible)	N
White/Blue (Convertible)	Q
White/Smoke (Convertible)	U

1987 Mustang Facts

Entering the tenth year with the same basic platform, Ford gave the Mustang a much needed facelift, emulating the aero-look of the company's other offerings. From the rear or from the side, the Mustang still looked like the same old Mustang, but at least from the front, it looked different.

The new GT achieved a totally different look through a redesigned nose with the flush fitting headlights. A ground effect skirt package with scoops on front of each wheel opening, extended around to the rear giving the car a much lower appearance. On the hatch sat a large rear wing, but the most noticeable change was the unusual louvered rear taillights.

The redesigned dash gave the Mustang a more modern look. Instrumentation was complete: fuel, water temperature, oil pressure, voltmeter, tachometer and a speedometer which still read only to 85 mph. The GTs got the articulated seats with power lumbar support and adjustable under-thigh support. Air conditioning controls, also new, were relocated in the center console top above the radio.

The standard engine on all LX Mustangs was still the 2.3L. The only optional engine (as the 3.8L V-6 was dropped) was the 5.0L HO. Thanks to a larger throttle body and better flowing cylinder heads derived from the 5.0 used on Ford trucks, the HO pumped out a very strong 225 hp (220 with the AOD) with 300 pounds-feet of torque.

The five-speed manual transmission was standard equipment across the board. With the 5.0L, axle ratios were unchanged from before, 2.73:1 with the five-speed or 3.08:1 with the optional automatic overdrive.

The suspension was also improved with many parts once used on the SVO. Increased wheel travel coupled with alignment changes gave the Mustang a better feel. Steering ratios, 20:1 for the LX and 14.7:1 on the GT and 5.0L-equipped LXs, was unchanged from 1986. Rear stabilizer bar size was upgraded to .82 inch on the 5.0L.

Front disc brakes were increased to 10.9 inches diameter for better braking, though the rear brakes were still 9-inch drums.

The GTs came with a new aluminum turbine wheel. The ten-hole aluminum wheels used on the 1986 GTs now became the standard wheel on Mustangs equipped with the 5.0L HO engine package.

The tilt-wheel was standard on the GTs. A new option for 1987 was the graphic equalizer. The only optional radio was the electronic AM/FM cassette stereo unit.

A lower titanium paint accent treatment was a no-cost option on the GT.

Leather was optional only on the LX and GT convertibles.

1987 LX convertible Ford Motor Co.

1987 GT

Chapter 24

1988 Mustang

Production Figures

66B 2dr Sedan	53,221	61B 3dr Hatchback	125,930
66B 2dr Convertible	32,074	Total	211,225

Serial Numbers

1FABP40A6JF000001
1FA — Ford Motor Co.
B — Restraint system (B-active belts)
P — Passenger car
40 — Body code (40-2dr LX, 41-3dr LX, 42-3dr GT, 44-2dr LX convertible, 45-2dr GT convertible)
A — Engine code
6 — Check digit which varies
J — Year (J-1988)
F — Plant (F-Dearborn)
000001 — Consecutive unit number

Location

Stamped on riveted plate on driver's side of dash, visible through the windshield; certification label attached to rear face of driver's door.

Engine Codes
A — 2.3L 1V 4 cyl 88 hp
M — 5.0L EFI V-8 225 hp(HO)

1988 Mustang Prices

	Retail
2dr LX Sedan, P40	$8,835.00
2dr LX Hatchback, P41	9,341.00
2dr LX Convertible, P44	13,702.00
2dr GT Hatchback, P42	12,745.00
2dr GT Convertible, P45	16,610.00
5.0L EFI HO 8 cyl (std. GT)	2,007.00
Automatic overdrive	
LX	515.00
GT	515.00
Tires, extra charge for four P195/75Rx14 WSW	82.00
Air conditioner, manual	788.00
Bracket, front license plate	N/C
Defroster, rear window	145.00
Heater, engine block immersion	18.00
Lock Group, power	237.00
Mirrors, dual electric remote	60.00
Mirrors, dual illuminated visor	100.00
Molding, bodyside insert stripe	49.00

Radio, electronic AM/FM with cassette tape	137.00
Sound system, premium	168.00
Radio credit option	(206.00)
Graphic equalizer	218.00
Roof, flip-up open air	355.00
Speed control	182.00
Steering wheel, tilt	124.00
Wheel covers, four wire style	178.00
Wheels, four styled road	178.00
Windows, power side	222.00
Emission system, Calif.	99.00
Emission system, high altitude	N/C
Leather articulated sport seats	
LX Convertible	780.00
GT Convertible	415.00
Vinyl seat trim	37.00
Lower titanium accent treatment (GT only)	N/C

1988 Exterior Colors	Code
Black	1C
Dark Grey Metallic	1D
Light Grey	1K
Bright Red	2I
Cabernet Red	2H
Medium Shadow Blue Metallic	3R
Tropical Yellow	66
Deep Shadow Blue Metallic	7N
Bright Regatta Blue Metallic	7H
Almond	6V
Oxford White	9L

1988 Interior Trim	Code
Regatta Blue	B
Scarlet	D
Smoke	G
Sand Beige	Y
White/Scarlet (Convertible)	N
White/Blue (Convertible)	Q
White/Smoke (Convertible)	U

1988 Mustang Facts

No significant changes were made on the 1988 Mustang.

Although the T-roof was discontinued after the 1987 model year, a few early 1988s were built with it.

Chapter 25

1989 Mustang

Production Figures
66B 2dr Sedan	50,560	61B 3dr Hatchback	116,965
66B 2dr Convertible	42,244	Total	209,769

Serial Numbers
1FABP40A6KF000001
1FA — Ford Motor Co.
B — Restraint system (B-active belts)
P — Passenger car
40 — Body code (40-2dr LX, 41-3dr LX, 42-3dr GT, 44-2dr LX convertible, 45-2dr GT convertible)
A — Engine code
6 — Check digit which varies
K — Year (K-1989)
F — Assembly plant (F-Dearborn)
000001 — Consecutive unit number

Location
Stamped on riveted plate on driver's side of dash, visible through the windshield; certification label attached on rear face of driver's door.

Engine Codes
A — 2.3L 1V 4 cyl 88 hp
M — 5.0L EFI V-8 225 hp (HO)

1989 Mustang Prices

	Retail
2dr LX Sedan, P40	$9,050.00
2dr LX Hatchback, P41	9,556.00
2dr LX Convertible, P44	14,140.00
2dr LX 5.0L Sport Sedan, P40	11,410.00
2dr LX 5.0L Sport Hatchback, P41	12,265.00
2dr LX 5.0L Sport Convertible, P44	17,001.00
2dr GT Hatchback, P42	13,272.00
2dr GT Convertible, P45	17,512.00
Automatic overdrive	
LX	515.00
GT	515.00
Tires, extra charge for four P195/75Rx14 WSW	82.00
Air conditioner, manual	807.00
Bracket, front license plate	N/C
Defroster, rear window	150.00
Heater, engine block immersion	20.00
Lock Group, power	246.00

Mirrors, dual electric remote	70.00
Mirrors, dual illuminated visor	100.00
Molding, bodyside insert stripe	61.00
Radio, electronic AM/FM with cassette tape & clock	137.00
Sound system, premium	168.00
Radio credit option	(245.00-382.00)*
Roof, flip-up open air	355.00
Speed control	191.00
Steering wheel, tilt	124.00
Wheel covers, four wire style	193.00
Wheels, four styled road	193.00
Windows, power side	232.00
Emission system, Calif.	100.00
Emission system, high altitude	N/C
Leather articulated sport seats	
LX Convertible	855.00
LX 5.0L Sport Convertible or GT Convertible	489.00
Vinyl seat trim	37.00
Lower titanium accent treatment (GT only)	N/C

*Depending on equipment package

1989 Exterior Colors

Color	Code
Black	1C
Dark Grey Metallic	1D
Light Grey	1K
Bright Red	21
Cabernet Red	2H
Medium Shadow Blue Metallic	3R
Tropical Yellow	66
Almond	6V
Bright Regatta Blue Metallic	7H
Deep Shadow Blue Metallic	7N
Oxford White	9L

1989 Interior Trim

Trim	Code
Regatta Blue	B
Scarlet	D
Smoke	G
Sand Beige	Y
White/Scarlet (Convertible)	N
White/Blue (Convertible)	Q
White/Smoke (Convertible)	U

1989 Mustang Facts

No significant changes were made on the 1989 Mustang.

The graphic equalizer was not available in 1989.

A clock function was added to the optional electronic AM/FM cassette stereo.

The LX models with the optional 5.0L were renamed LX 5.0L Sport. During the model year, a 140 mph speedometer took the place of the standard 85 mph unit on 5.0L powered Mustangs.

Chapter 26

1990 Mustang

Production Figures

66B 2dr Sedan	22,503
66B 2dr Convertible	26,958
61B 3dr Hatchback	78,728
Total	128,189
Mustang GT	34,435
Mustang LX 5.0l	44,851

Serial Numbers
1FACP40A6LF000001
1FA — Ford Motor Co.
C — Restraint system (C-air bags & active belts)
P — Passenger car
40 — Body code (40-2dr LX, 41-3dr LX, 42-3dr GT, 44-2dr convertible, 45-GT convertible)
A — Engine code
6 — Check digit which varies
L — Year (L-1990)
F — Assembly plant (F-Dearborn)
000001 — Consecutive unit number

Location
Stamped on riveted plate on driver's side of dash, visible through the windshield; certification label attached on rear face of driver's door.

Engine Codes
A — 2.3L EFI 4 cyl 88 hp
M — 5.0L EFI V-8 225 hp(HO)

1990 Mustang Prices

	Retail
2dr LX Sedan, P40	$9,753.00
2dr LX Hatchback, P41	10,259.00
2dr LX Convertible, P44	14,810.00
2dr LX 5.0L Sport Sedan, P40	12,222.00
2dr LX 5.0L Sport Hatchback, P41	13,065.00
2dr LX 5.0L Sport Convertible, P44	17,796.00
2dr GT Hatchback, P42	14,044.00
2dr GT Convertible, P45	18,418.00
Automatic overdrive	539.00
Tires, extra charge for four P195/75Rx24 WSW on LX	82.00

Air conditioner, manual control	807.00
Defroster, rear window	150.00
Heater, engine block immersion	20.00
Mirrors, dual illuminated visor	100.00
Power equipment group (std. on convertibles)	507.00
Roof, flip-up open air	355.00
Radio, electronic AM/FM with cassette & clock	137.00
Sound system, premium	168.00
Radio credit option*	(245.00-550.00)
Speed control	191.00
Wheel covers, wire style	193.00
Emissions system, Calif.	100.00
Paint, clearcoat	91.00
Lower titanium accent treatment (GT only)	159.00
Leather seating surface, articulated sport seats	489.00
Vinyl seat trim	37.00

*Depending on option package

1990 Exterior Colors

Color	Code
Cabernet Red	EH
Bright Red	EP
Black	YC
Oxford White	YO
Bright Yellow	AG
Wild Strawberry	EL
Crystal Blue	MA
Twilight Blue	MK
Deep Emerald Green	YF*
Deep Titanium	YU
Light Titanium	YF

*On LX 5.0L Convertible Special only

1990 Interior Trim

Trim	Code
Titanium	A
Crystal Blue	B
Scarlet	D
Ebony	J
White/Titanium	L
White/Scarlet	N
White/Crystal Blue	W
White	

Convertible Top Colors

Black
Dark Blue
White

1990 Mustang Facts

Again, no significant styling or mechanical changes were made on the 1990 Mustangs, save for the driver's side air bag.

Interior changes included the deletion of the center console armrest, the addition of door trim map pockets and a driver's footrest on the LX.

The standard wheel covers are a finned design, with the wire wheel covers optional on the LX.

The LX 5.0L and GT Mustangs got 140 mph speedometers.

A limited edition convertible, 2,000 units, for 1990½ is planned. It will come in deep emerald green clearcoat metallic exterior paint and a white interior.

Chapter 27

1991 Mustang

Production Figures

66B 2dr Sedan	19,447
66B 2dr Convertible	21,513
61B 3dr Hatchback	57,777
Total	98,737
Mustang GT	24,428
Mustang LX 5.0l	27,880

Serial Numbers
1FACP40E6MF000001
1FA — Ford Motor Co.
C — Restraint system (C-Air bags & active belts)
P — Passenger car
40 — Body code (40-2dr LX, 41-3dr LX, 42-3dr GT, 44-2dr convertible, 45-GT
Convertible)
E — Engine code
6 — Check digit which varies
M — Model year (M-1991)
F — Assembly plant (F-Dearborn)
000001 — Consecutive unit number assembly number

Location
Stamped on riveted plate on driver's side of dash, visible through the windshield. Certification Label attached on rear face of driver's door.

Engine codes
S — 2.3L EFI four cylinder, 105hp.
E — 5.0L EFI V-8, 225hp (HO)

1991 Mustang Prices*

LX, Sedan P40	$10,702
LX, Hatchback P41	11,208
LX, Convertible P44	16,767
LX 5.0l, Sedan	13,815
LX 5.0l, Hatchback	14,600
LX 5.0l, Convertible	19,787
GT, Hatchback	15,579
GT, Convertible P45	20,409
44L 4-Speed Automatic Overdrive	595
422 California Emission System	100
428 High Altitude Emission System	N/C
Vinyl Low Back seats	37
Leather Seats	499
45C Axle, Optional Ratio (N/A LX)	N/C
18C Cargo Tie-Down Net	66

60A Custom Equipment Group, includes Manual A/C, Dual Illum. Mirrors	
LX Sedan/Hatchback	917
All Others	817
57Q Defroster, Rear Window	160
12H Floor Mats, Front	33
915 Graphic Equalizer, w/Premium Sound	139
wo/Premium Sound	307
Paint, Clearcoat	91
913 Sound System, Premium	168
589 Radio, Electronic AM/FM w/Cassette	155
58Y Radio Credit Option	(245-568)
13C Roof, Flip-Up Open Air (Hatchback)	355
525 Speed Control	210
954 Titanium Lower Bodyside Accent Treatment (GT Only)	159
644 Wheel Covers, Wire Style	N/C
64W Wheels, Cast Aluminum	167-360
64J Wheels, Styled Road (LX Only)	193
153 Bracket, Front License Plate	N/C
41H Heater, Engine Block Immersion	20

*As of September 27, 1990

1991 Exterior Colors   Code

Medium Red	EM
Bright Red	EP
Black	YD
Oxford White	YO
Wild Strawberry	EL
Light Crystal Blue	MA
Twilight Blue	MK
Deep Emerald Green	YF
Medium Titanium	YG
Titanium Frost	YX

1991 Convertible Top Colors

Black	AA
Dark Blue	RR
White	WW

1991 Interior Trim

Titanium	BA (Vinyl)		FA (Cloth)		DA (Cloth/Vinyl)		CA (Leather)	
Crystal Blue	BB	"	FB	"				
Scarlet Red	BD	"	FB	"	DD	" "		
Black	BJ	"	FJ	"	DJ	" "	CJ	"
White/Titanium*	BL	"					CL	"
White/Red*	BN	"					CN	"

*Convertible only

1991 Mustang Facts

The Mustang continued unchanged for 1991. New to the option list were the front Floor mats and Cargo Tie-Down Net.

Chapter 28

1992 Mustang

Production Figures
66B 2dr Sedan	15,717
66B 2dr Convertible	23,470
61B 3dr Hatchback	40,093
Total	79,280
Mustang GT	20,445
Mustang LX 5.0l	19,131

Serial Numbers
1FACP40E6NF000001
1FA — Ford Motor Co.
C — Restraint system (C-Air bags & active belts)
P — Passenger car
40 — Body code (40-2dr LX, 41-3dr LX, 42-3dr GT, 44-2dr convertible, 45-GT
Convertible)
E — Engine code
6 — Check digit which varies
N — Model year (N-1992)
F — Assembly plant (F-Dearborn)
000001 — Consecutive Unit Number

Location:
Stamped on riveted plate on driver's side of dash, visible through the windshield. Certification Label attached on rear face of driver's door.

Engine codes
S — 2.3L EFI four cylinder, 105hp.
E — 5.0L EFI V-8, 225hp (HO)

1992 Mustang Prices
LX, Sedan P40	$10,125
LX, Hatchback P41	10,721
LX, Convertible P44	16,899
LX 5.0l, Sedan	13,422
LX 5.0l, Hatchback	14,207
LX 5.0l, Convertible	19,644
GT, Hatchback	15,243
GT, Convertible P45	20,199
44L 4-Speed Automatic Overdrive	595
Vinyl Low Back seats	76
Leather Seats	523

45C Axle, Optional Ratio (N/A LX)	N/C
572 Air Conditioning, Manual	817
60K Convenience Group	99
57Q Defroster, Rear Window	170
915 Graphic Equalizer, w/Premium Sound	139
w/o Premium Sound	307
677 Mirrors, Dual Illuminated Visor 100	
Paint, Clearcoat	91
61A Power Equipment Group (Std Conv.)	604
913 Sound System, Premium	168
589 Radio, Electronic AM/FM w/Cassette	155
58Y Radio Credit Option	(245-568)
13C Roof, Flip-Up Open Air (Hatchback)	355
217 Seat, 4-Way Power Driver's	183
525 Speed Control	224
954 Titanium Lower Bodyside Accent	
Treatment (GT Only)	159
64W Wheels, Cast Aluminum	208-401
64J Wheels, Styled Road (LX Only)	193
153 Bracket, Front License Plate	N/C
41H Heater, Engine Block Immersion	20

1992 Exterior Colors

	Codes	Convertible Tops	
Medium Red	EM	Black	AA
Bright Red	EP	Dark Blue	RR
Black	UA	White	WW
Oxford White	YO		
Wild Strawberry	EL		
Bimini Blue	K3		
Twilight Blue	MK		
Ultra Blue	MM		
Deep Emerald Green	PA		
Medium Titanium	YG		
Titanium Frost	YX		

1992 Interior Trim

Titanium	BA	(Vinyl)	FA	(Cloth)	DA	(Cloth/Vinyl)	CA	(Leather)	
Crystal Blue	BB	"	FB	"					
Scarlet Red	BD	"	FB	"	DD	"	"		
Black	BJ	"	FJ	"	DJ	"	"	CJ	"
White/Titanium*	BL	"						CL	"
White/Red*	BN	"						CN	"

*Convertible only

1992 Mustang Facts

The Mustang continued unchanged. 2,019 Vibrant Red 5.0l LX Convertibles were sold as 1992-1/2 models. These featured special Vibrant Red paint schemes.

Chapter 29

1993 Mustang

Production
66B 2dr Sedan	24,851
66B 2dr Convertible	27,300
61B 3dr Hatchback	62,077
Total	114,228
Mustang GT	26,101
Mustang LX 5.0l	22,902

Serial Numbers
1FACP40E6PF000001
1FA — Ford Motor Co.
C — Restraint system (C-Air bags & active belts)
P — Passenger car
40 — Body code (40-2dr LX, 41-3dr LX, 42-3dr GT, 44-2dr convertible, 45-GT
Convertible)
E — Engine code
6 — Check digit which varies
P — Model year (P-1993)
F — Assembly plant (F-Dearborn)
000001 — Consecutive Unit Number

Location:
Stamped on riveted plate on driver's side of dash, visible through the windshield. Certification Label attached on rear face of driver's door.

Engine codes
S — 2.3L EFI four cylinder, 105hp.
E — 5.0L EFI V-8, 205hp (HO)
D — 5.0L Cobra EFI 235hp

1993 Mustang Prices*
LX, Sedan P40	$11,159
LX, Hatchback P41	11,664
LX, Convertible P44	17,988
LX 5.0l, Sedan	14,366
LX 5.0l, Hatchback	15,150
LX 5.0l, Convertible	20,733
GT, Hatchback	16,187
GT, Convertible P45	21,288
44L 4-Speed Automatic Overdrive	595

422 California Emissions System	100
Vinyl Low Back seats	76
Leather Seats	523
45C Axle, Optional Ratio (N/A LX)	N/C
672 Air Conditioning, Manual	817
60K Convenience Group	99
57Q Defroster, Rear Window	170
677 Mirrors, Dual Illuminated Visor	100
Paint, Clearcoat	91
61A Power Equipment Group(Std. Conv.)	604
585 Radio, Electronic AM/FM w/CD Player	629
588 Radio, Electronic Premium Cassette with Premium Sound	339
58Y Radio Credit Option	(245-584)
13C Roof, Flip-Up Open Air (Hatchback)	355
217 Seat, 4-Way Power Driver's	183
525 Speed Control	224
64W Wheels, Cast Aluminum	208-401
64J Wheels, Styled Road (LX Only)	193
153 Bracket, Front License Plate	N/C
41H Heater, Engine Block Immersion	20

As of July 15, 1992

1993 Exterior Colors

	Code
Bright Red	EP
Black	YC
Vibrant White	WB
Electric Red	EG
Bright Blue	KF
Royal Blue	LA
Reef Blue	PD
Bright Calypso Green	PM
Silver	YN

Convertible tops

Black	AA
Dark Blue	RR
White	WW

1993 Interior Trim

Opal Gray	BA (Vinyl)	FA (Cloth)	DA (Cloth/Vinyl)	CA (Leather)
Crystal Blue	BB "	FB "		
Ruby Red	BD "	FB "	DD " "	
Black	BJ "	FJ "	DJ " "	CJ "
White/Opal Gray*	BL "			CL "
White/Ruby Red*	BN "			CN "

*Convertible only

1993 Mustang Facts

New for 1993 was the CD Player option. There were 1,500 White Limited Edition LX 5.0l Convertibles and 1,503 Yellow

The 1993 Mustang Cobra stands out as high point among third generation performance Mustangs. Ford Motor Co.

Limited Edition LX 5.0l Convertibles. Both were released as 1993-1/3 models.

For 1993 the 5.0l HO V-8 was downrated to 205hp. The reduction was partly due to a change in the way Ford measured engine horsepower and also due to the cumulative effect of minor mechanical changes to the engine since 1987. A restrictive resonator in the inlet air tract reduced power by 5-7hp, a revised camshaft in 1989 accounted for three hp, the change over to a Mass Air System in 1989 from the previous Speed Density accounts for 2-3hp. Minor changes in the exhaust system also helped to reduce power as well. The big news for 1993 was the introduction of the Cobra. Visually, the Cobra got a side skirt treatment, a redesigned rear wing spoiler and appropriate Cobra badges on the front fenders and grille. The Cobra's 5.0l V-8 was rated at 235hp, due primarily to better flowing GT-40 cylinder heads. The suspension also differed from the regular Mustang GT. Instead of the GT's springs, the Cobra used the softer rear springs from the base 2.3l powered Mustangs along with a smaller rear anti-sway bar. In the front, the Cobra was equipped with the 2.3l struts. The Cobra, though, did come with wider 245/45ZR Goodyear tires mounted on unique 17in diameter wheels. The Cobra was designed to handle better than the regular GT and at the same time provide a more comfortable ride. 4,993 Cobras were built.

There were also 107 Cobra R Mustangs built in 1993 by Ford's SVT Division. Mechanically similar to the street Cobra, the R model was designed for use in showroom stock racing. Rather than using the soft Cobra suspension, the R model Mustangs got much firmer springs and shocks/struts and all were painted white.

Chapter 30

1994 Mustang

Production Figures
2dr Coupe 3.8l	42,883
2dr Convertible 3.8l	18,333
2dr Coupe GT	30,592
2dr Convertible GT	25,381
2dr Coupe Cobra	5,009
2dr Convertible Cobra	10,000
Total Mustang	123,198

Serial Numbers
1FALP4046RF000001
1FA — Ford Motor Co.
L — Restraint system (L-Air bags & active belts)
P — Passenger car
40 — Body code (40-2dr, 42-2dr GT, 44-2dr convertible, 45-GT Convertible)
4 — Engine code
6 — Check digit which varies
R — Model year (R-1994)
F — Assembly plant (F-Dearborn)
000001 — Consecutive Unit Number

Location:
Stamped on riveted plate on driver's side of dash, visible through the windshield. Certification Label attached on rear face of driver's door.

Engine Codes
4 — 3.8l EFI V-6, 145hp
E — 5.0l EFI V-8, 215hp
0 — 5.0l EFI V-8, 240hp

1994 Mustang Prices
Mustang Coupe, P40	$13,365
Mustang Convertible, P44	20,160
Mustang GT Coupe, P42	17,280
Mustang GT Convertible, P45	21,790
Mustang Cobra Coupe, P42	21,300**
Mustang Cobra Convertible, P45	25,605**
44P 4-Speed Automatic Overdrive	790
422 California Emissions System	95
Leather Seats	500
672 Air Conditioning, Manual	780
552 Anti-Lock Braking System	565

18A Anti-Theft System	235
45C Axle, Optional Ratio (GT only)	N/C
153 Bracket, Front License Plate	N/C
13T Convertible Hardtop	1,545
57Q Defroster, Rear Window	160
12H Floor Mats, Front	30
41H Heater, Engine Block Immersion	20
61A Group 1 Includes:	505
Power Side Windows	
Power Door Locks	
Power Decklid Release	
63A Group 2 Includes:	
Speed Control Mirrors, Dual Illuminated Visors, Radio, Electronic AM/FM Stereo/Cassette/Premium Sound	
w/Mustang Coupe	870
w/Mustang Convertible	775
w/GT Coupe/Convertible	510
60K Group 3 Includes:	310
Remote Keyless/Illuminated Entry, Cargo Net	
961 Moldings, Bodyside	50
21Y Power Driver's Seat Credit	(135)
58M Radio, Electronic AM/FM Stereo Cassette	165
586 Radio, Mach 460 Electronic AM/FM Stereo Cassette	
w/Group 2	375
w/o Group 2	670
917 Compact Disc Player	475
64J Wheels, 15in Cast Aluminum	265
64H Wheels, Unique 17in (GT Only)	380

*As of October 15, 1993
**As of January 12, 1995

1994 Exterior Colors

	Code
Canary Yellow	BZ*
Rio Red	E8
Laser Red	E9
Vibrant Red	ES*
Iris	GC
Bright Blue	KF
Deep Forest Green	NA
Teal	RD
Black	UA
Opal Frost	WJ
Crystal White	ZF

*GT only

1994 Convertible Top Colors

Black	AA
Saddle	MM
White	WW

1994 Interior Trim

	Bright Red	Saddle	Opal Gray	Black	White*
Coupe & Convertible					
Cloth	6F	6S	66	6J	-
Leather	-	-	-	-	Z
GT Coupe & Convertible					
Cloth	2F	2S	26	2J	-
Leather	4F	4S	46	4J	Z

*Convertible only

1994 Mustang Facts

For 1994 Mustang finally got a new body—a two-door Coupe and Convertible built on an improved version of the Fox platform. Wheelbase was longer by .75in while track is 3.7in wider on the base Mustang and 1.9in longer on the GT. Four-wheel disc brakes were standard with ABS optional. Dual airbags were standard, as was a power driver's seat and a tilt steering wheel.

The suspension was basically carry over as well—MacPherson struts in the front with a live axle in the rear. The quad-shock arrangement is standard on the GT. One improvement was the use of a rear anti-sway bar on the base Mustang. 15x6.5in wheels are standard, 16.75in wheels are standard on the GT with 17x8in wheels optional on the GT. The optional GT wheels feature P245/45ZR17 tires.

Although the GT's engine is uprated to 215hp for 1994, the big change was on the base Mustang. The anemic 2.3l four cylinder was replaced by the 3.8l V-6, which was the same engine used last on 1986 Mustangs. With an updated tuned port injection system and tubular headers, the V-6 put out a respectable 145hp.

Standard on both engines is the Borg Warner T-5 five-speed manual transmission with the four-speed automatic AOD optional.

Although the optional hardtop for the convertible was listed as an option, it did not actually become available until the 1995 model year.

The Cobra Mustang was available on the Coupe and Convertible. By using the GT-40 cylinder heads, a special intake manifold, a revised camshaft profile, higher lift 1.72:1 Crane rocker arms and higher rate fuel injectors, power on the Cobra 5.0l is 240hp.

The Cobra's suspension uses the base V-6 springs along with a smaller front 0.98in anti-sway bar (1.18in/GT, 1.06in/base) and a larger rear one measuring 1.06in(.94in/GT, .83in/base) The Cobra does have superior brakes as it employs 13.0in front/11.65in rear PBR brakes with ABS.

Visually the Cobra uses a different front fascia, rear wing and wheels. In the interior, the Cobra got appropriate identification along with white gauges with black markings; the GT has white on black. All standard power accessories and A/C are standard with the Cobra.

Color availability on the Cobra is limited to Crystal White, Black, or Rio Red.

The new Mustang was chosen to be the 1994 Indy 500 Pace Car.

The 1994 GT retains its Mustang heritage with three-bar taillamps and a pony on the grill. Ford Motor Co.

The standard engine for the 1994 Mustang was Ford's 3.8L V-6. Ford Motor Co.

Chapter 31

1995 Mustang

Production Figures
2dr Sedan	137,722
2dr Convertible	48,264
Total	185,986

Serial Numbers
1FALP4046SF000001
1FA — Ford Motor Co.
L — Restraint system (Air bags & active belts)
P — Passenger car
40 — Body code (40-2dr, 42-2dr GT, 44-2dr convertible, 45-GT Convertible)
4 — Engine code
6 — Check digit which varies
S — Model year (S-1995)
F — Assembly plant (F-Dearborn)
000001 — Consecutive unit number

Location
Stamped on riveted plate on driver's side of dash, visible through the windshield. Certification Label attached on rear face of driver's door.

Engine codes
4 — 3.8L
E — 5.0L EFI V-8, 215hp
0 — 5.0l EFI V-8, 240hp
    5.8l EFI V-8, 300hp

1995 Mustang Prices*
Mustang Coupe, P40	$14,530
Mustang Convertible, P44	20,995
Mustang GTS Coupe, P42	16,910
Mustang GT Coupe, P42	18,105
Mustang GT Convertible, P45	22,795
Mustang Cobra Coupe, P42	21,300
Mustang Cobra Convertible, P45	25,605
44P 4-Speed Automatic Overdrive	815
422 California Emissions System	95
Leather Seats	500
217 Power Driver's Seat	175
672 Air Conditioning, Manual	855
552 Anti-Lock Braking System	565
18A Anti-Theft System	145
45C Axle, Optional Ratio (GT/GTS only)	N/C

153 Bracket, Front License Plate	N/C
13T Convertible Hardtop	1,825
57Q Defroster, Rear Window	160
12H Floor Mats, Front	30
41H Heater, Engine Block Immersion	20
61A Group 1 Includes:	505
Power Side Windows, Power Door Locks, Power Decklid Release	
63A Group 2 Includes:	
Speed Control, Mirrors, Dual Illuminated Visors, Radio, Electronic AM/FM Stereo/Cassette/Premium Sound, Wheels, 15" Cast Aluminum (NA w/GT)	
w/Mustang Coupe	870
w/Mustang Convertible	775
w/GT Coupe/Convertible	510
60K Group 3 Includes:	310
Remote Keyless/Illuminates Entry, Cargo Net	
961 Moldings, Bodyside	50
21Y Power Driver's Seat Credit	(135)
58M Radio, Electronic AM/FM Stereo Cassette	165
586 Radio, Mach 460 Electronic AM/FM Stereo Cassette	
w/Group 2	375
w/o Group 2	670
585 Compact Disc Radio/Premium Sound	
w/Group 2	140
w/58M	270
w/58F	435
912 Compact Disc Radio/Mach 460(w/585)	375
917 Compact Disc Player	475
64J Wheels, 15" Cast Aluminum	265
64H Wheels, Unique 17" (GT Only)	380

* as of January 12. 1995

1995 Exterior Colors   Codes

Canary Yellow	BZ*
Rio Red	E8
Laser Red	E9
Vibrant Red	ES
Sapphire Blue	JA*
Bright Blue	KF
Deep Forest Green	NA
Teal	RD
Black	UA
Opal Frost	WJ
Crystal White	ZR

*GT/GTS only

1995 Convertible Top Colors

Black	AA
Saddle	MM
White	WW

1995 Interior Trim

	Bright Red	Saddle	Opal Gray	Black	White*
Coupe & Convertible					
Cloth	6F	6S	66	6J	-
Leather	-	-	-	-	1Z
GT Coupe & Convertible					
Cloth	2F	2S	26	2J	-
Leather	4F	4S	46	4J	4Z
GTS Coupe					
Cloth	6F	S	66	6J	-

*Convertible only

1995 Mustang Facts

The 1995 Mustang was a carry over model for 1995. The optional Convertible hardtop finally became available in 1995.

The GTS Mustang was the GT model without some of the GT features such as the sport seats, rear spoiler and fog lamps. 1995 was the last year for the 5.0l V-8. For 1996 and later, the 4.6l Modular V-8 was the top engine option.

Ford's SVT (Special Vehicles Team) built 250 Cobra R Mustangs for showroom stock racing in 1995. Unlike the previous 1993 R model, the 1995 R was equipped with Ford's 5.8l (351ci) V-8 rated at 300hp with 365 lbs/ft torque. As the T-5 manual transmission was not designed for such torque, SVT used the stronger Tremec unit.

The suspension on the Cobra R model featured heavier front/rear springs with larger front/rear anti-sway bars and Koni shocks and struts. The Cobra R used the same PBR brakes that are standard on the street Cobra with larger 17x9in wheels mounting P255/45ZR-17 B.F. Goodrich Comp T/A tires. The Cobra R was also equipped with a larger 20gal gas tank (vs 15.4 stock) and a unique fiberglass hood. In keeping with its performance mission, the R did not have a rear seat, radio, A/C, power windows/locks/mirrors, sound insulation or fog lights. The only color available was white.

250 Cobra R models were built.

The Mustang Cobra R shares body panels with the Mustang Cobra, but has a unique hood and wheels. All were painted white. Ford Motor Co.

Chapter 32

1996 Mustang

Production Figures

Base Coupe	61,187
Base Convertible	15,246
GT Coupe	31,624
GT Convertible	17,917
Cobra Coupe	7,496
Cobra Convertible	2,510
Total	135,620

Serial Numbers

1FALP45W6TF000001
1FA — Ford Motor Co.
L — Restraint system (L-Air bags & active belts)
P — Passenger car
45 — Body code (40-coupe, 42-GT coupe, 44-convertible, 45-GT convertible, 47-Cobra coupe, 46-Cobra convertible)
W — Engine code
6 — Check digit which varies
T — Model year (T-1996)
F — Assembly plant (F-Dearborn)
000001 — Consecutive unit number

Location:

Stamped on riveted plate on driver's side of dash, visible through the windshield. Certification label attached on rear face of driver's door.

Engine Codes

4 — 3.8l EFI V-6 150 hp
W — 4.6l SOHC EFI V-8 215 hp
V — 4.6l DOHC EFI V-8 305 hp (SVT Cobra)

1996 Mustang Prices*

Coupe, P40	$15,180
Convertible, P44	21,060
GT Coupe, P42	17,610
GT Convertible, P45	23,495
Cobra Coupe, P47	24,810
Cobra Convertible, P46	27,580
44U 4-speed automatic overdrive	815
61A Option Group 1 includes:	
Power side windows/door locks, power deck lid release	505
63A Option Group 2 includes:	
Speed control, 15-in. aluminum wheels, AM/FM ETR w/cass. & premium sound	
Base	775
GT	510
65A Option Group 3 includes:	

Fog lamps, GT sport seats, leather-wrapped steering wheel, rear spoiler, dual illum. visors	690
572 Air conditioning, manual	895
18A Anti-theft system	145
45C Axle, optional ratio (3.27 GT only)	200
552 Brakes, anti-lock	570
57Q Defroster, rear window	170
422 Emission system, Calif.	100
428 Emission system, high altitude	N/C
12H Floor mats, front	30
41H Heater, engine block immersion	20
143 Keyless entry	270
677 Mirrors, dual illum.	95
961 Molding, bodyside	60
58M Radio, electronic AM/FM stereo cassette & premium sound	165
58H Radio, electronic AM/FM stereo cassette & premium sound	130–295
586 Radio, Mach 460 electronic AM/FM stereo cassette	395–690
917 Compact disc player	295
534 Rear spoiler	195
217 Power driver's seat	175
64J Wheels, 15-in cast aluminum (base only)	265
64Y Wheels, Unique 17-in (GT only)	400
Leather seats	500

* as of March 7, 1996

1996 Exterior Colors

	Codes
Bright Tangerine	CM
Rio Red	E8
Laser Red	E9
Deep Violet	JU
Moonlight Blue	KM
Mystic	LF (SVT Cobra only)
Deep Forest Green	NA
Pacific Green	PS
Black	UA
Opal Frost	WJ
Crystal White	ZR

Convertible Tops

Black	A
Saddle	M
White	W

1996 Interior Trim

	Saddle	Opal Gray	Black	White
Coupe & Convertible				
Cloth	6S	66	6J	-
Leather	-	-	-	1Z
GT Coupe & Convertible				
Cloth	7J	26	42	-
Leather	4S	46	4J	4Z

1996 Mustang Facts

Although the SN-95 platform was carried over with minimal changes, there were significant changes under the hood.

The 5.0l V-8 was finally retired and replaced by a 4.6l version of the Ford's modular V-8. The most unique feature of the modular V-8 was its chain-driven Single Overhead Camshaft Design. The 4.6l features considerably improved porting over the old 5.0l and thus has better breathing potential. The engine used a 65-mm throttle body with an 80-mm Mass Air Sensor. It was rated at 215 hp at 4,400 rpm with 285 ft/lbs torque at 3,500 rpm.

Besides the SOHC cylinder heads, the engine featured four-bolt main bearing caps for durability and a relatively square bore and stroke, 3.55x3.54 inch for good low-end torque.

Along with the new engine, the 1996 V-8–powered Mustangs got a new transmission, the Borg-Warner T45 five-speed. The transmission, which weighs in at 110 pounds, had a torque rating of 320 ft/lbs. And unlike the T5, the bell housing was an integral part of the transmission casing, which makes for a more rigid structure.

The 8.8-inch rear axle was carried over from 1995 and used a 2.73:1 axle ratio. Optional was a 3.27:1 ratio.

The GT's suspension was carried over for 1996. In the front, 400/505 lb/in variable-rate springs were used with a 30-mm anti-sway bar. In the rear, 165-265 lb/in variable-rate springs with a rear anti-sway bar measuring 25 mm were used. The rear bar for 1996 was 1 mm larger than the one used in 1995.

The brakes on all Mustangs were upgraded to a four-wheel disc setup. The front discs measured 10.8 inches and the rears 10.5 inches. ABS was optional.

Externally, all Mustangs received a new taillamp treatment, featuring three vertical bars. Mustang GTs were also equipped with new "GT 4.6l" fender emblems.

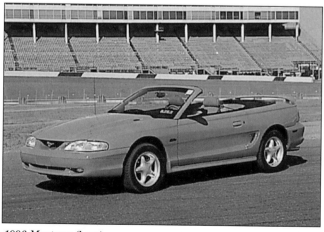

1996 Mustang (base)

Chapter 33

1997 Mustang

Production Figures

Base Coupe	56,812
Base Convertible	11,606
GT Coupe	18,464
GT Convertible	11,413
Cobra Coupe	6,961
Cobra Convertible	3,088
Total	108,344

Serial Numbers

1FALP45W6VF000001
1FA — Ford Motor Co.
L — Restraint system (L-Air bags & active belts)
P — Passenger car
45 — Body code (40-coupe, 42-GT coupe, 44-convertible, 45-GT convertible,
47-coupe, 46-convertible)
W — Engine code
6 — Check digit which varies
V — Model year (V-1997)
F — Assembly plant (F-Dearborn)
000001 — Consecutive unit number

Location:
Stamped on riveted plate on driver's side of dash, visible through the windshield. Certification label attached on rear face of driver's door.

Engine Codes
4 — 3.8l EFI V-6 150 hp
W — 4.6l SOHC EFI V-8 215 hp
V — 4.6l DOHC EFI V-8 305 hp (SVT Cobra)

1997 Mustang Prices*

Coupe, P40	$15,880
Convertible, P44	21,280
GT Coupe, P42	18,525
GT Convertible, P45	24,510
Cobra Coupe, P47	25,335
Cobra Convertible, P46	28,135
44U 4-speed automatic overdrive	815
54A Sport Appearance Group includes:	
Rear spoiler, 15-in polished wheels,	
leather-wrapped steering wheel,	
lower bodyside accent stripe	
Base	345
61A Option Group 1 includes:	
Power side windows/door locks,	

power deck lid release	565
63A Option Group 2 includes:	
Speed control, 15-in aluminum wheels, AM/FM ETR w/cass. & premium sound	
Base	775
GT	510
65A Option Group 3 includes:	
Fog lamps, GT sport seats, leather-wrapped steering wheel, rear spoiler, dual illum. visors	690
572 Air conditioning, manual	895
18A Anti-theft system	145
45C Axle, optional ratio (3.27 GT only)	200
153 Bracket, front license plate	N/C
552 Brakes, anti-lock	570
57Q Defroster, rear window	190
422 Emission system, Calif.	170
428 Emission system, high altitude	N/C
12H Floor mats, front	30
41H Heater, engine block immersion	20
143 Keyless entry	270
677 Mirrors, dual illum.	95
961 Molding, bodyside	60
217 Power driver's seat	210
58M Radio, electronic AM/FM stereo cassette & premium sound	165
586 Radio, Mach 460 electronic AM/FM stereo cassette	395–690
917 Compact disc player	295
534 Rear spoiler (Std. GT conv.)	195
525 Speed control	215
64J Wheels, 15-in cast aluminum (base only)	265
64Y Wheels, polished 17-in (GT only)	500
Leather seats	500

* as of March 12, 1997

1997 Exterior Colors

	Codes
Aztec Gold	AZ
Autumn Orange	BG
Rio Red	E8
Laser Red	E9
Deep Violet	JU
Moonlight Blue	KM
Deep Forest Green	NA
Pacific Green	PS
Black	UA
Crystal White	ZR

Convertible Tops

Black	A
Saddle	M
White	W

1997 Interior Trim

	Saddle	Black	Medium Graphite
Coupe & Convertible			
Cloth	ZS	ZJ	-
Leather	-	-	12
GT Coupe & Convertible			
Cloth	7J	42	72
Leather	4S	4J	42

1997 Mustang Facts

The 1997 Mustang line received minimal changes.

The Mustang GT got a new flecked seat fabric pattern and Medium Graphite replaced the previous white used in 1996. New 17-inch wheels became optional on the GT, which had a dark gray metallic center.

The PATS (Passive Anti-Theft System) previously used on the 1996 GT and Cobra models became standard on all 1997 Mustangs. The PATS system used an encoded ignition key with a transponder to electronically disable the engine if the transponder code does not match a preset code in the car's EEC-V control system.

All automatic-equipped Mustangs were fitted with a new, thicker shifter.

Save for a slight change in the front upper grille opening, there wasn't any change to the Mustang. This opening ducted more air to the new cross-the-board Mustang cooling system. All models had a wider and taller radiator and a larger diameter fan.

1997 Saleen S281

Chapter 34
1998 Mustang

Production Figures
Base Coupe	99,801
Base Convertible	21,254
GT Coupe	28,789
GT Convertible	17,024
Cobra Coupe	5,174
Cobra Convertible	3,480
Total	175,522

Serial Numbers
1FALP45W6WF000001
1FA — Ford Motor Co.
L — Restraint system (L-Air bags & active belts)
P — Passenger car
45 — Body code (40-coupe, 42-GT coupe, 44-convertible, 45-GT convertible,
47-Cobra coupe, 46-Cobra convertible)
W — Engine code
6 — Check digit which varies
W — Model year (W-1998)
F — Assembly plant (F-Dearborn)
000001 — Consecutive unit number

Location:
Stamped on riveted plate on driver's side of dash, visible through the windshield. Certification label attached on rear face of driver's door.

Engine Codes
4 — 3.8l EFI V-6 150 hp
W — 4.6l SOHC EFI V-8 225 hp
V — 4.6l DOHC EFI V-8 305 hp (SVT Cobra)

1998 Mustang Prices*
Coupe, P40	$16,150
Convertible, P44	20,650
GT Coupe, P42	20,150
GT Convertible, P45	24,150
Cobra Coupe, P47	25,710
Cobra Convertible, P46	28,510
44U 4-speed automatic overdrive	815
60C Convenience Group includes: Front floor mats, rear window defroster, speed control, power driver's seat	
Base	495
GT	295

54G GT Sport Group includes:
17-in 5-spoke aluminum wheels, hood stripe and wraparound fender stripes, leather-wrapped shift knob (manual trans. only), engine oil cooler 595
54V V6 Sport Appearance Group includes:
16-in cast aluminum wheels, rear spoiler, leather-wrapped steering wheel, lower bodyside accent stripe
Base 345
572 Air conditioning, manual 895
18A Anti-theft system 145
45C Axle, optional ratio (3.27 GT only) 200
552 Brakes, anti-lock 570
57Q Defroster, rear window 190
422 Emission system, Calif. 170
428 Emission system, high altitude N/C
41H Heater, engine block immersion 20
677 Mirrors, dual illum. visor 95
588 Radio, Mach 460 electronic AM/FM stereo cassette/equalizer 395
534 Rear spoiler (Std. GT) 195
Leather seats 500
64Y Wheels, polished 17-in (GT Only) 500
* as of March 15, 1998

1998 Exterior Colors — Codes

Color	Code
Chrome Yellow	BZ
Autumn Orange	BG
Rio Red	E8
Laser Red	E9
Performance Red	ES
Dark Green Satin	FU
Atlantic Blue	K6
Bright Atlantic Blu	K7
Pacific Green	PS
Black	UA
Silver	YN
Crystal White	ZR

Convertible Tops

Black	A
Saddle	M
White	W

1998 Interior Trim

	Saddle	Black	Medium Graphite
Coupe & Convertible			
Cloth	ZS	ZJ	-
Leather	-	-	12

GT Coupe & Convertible			
Cloth	7J	42	72
Leather	4S	J	42

1998 Mustang Facts

Once again, changes were minimal to the Mustang line. The 4.6l received minor tweaks to produce 225 hp. The optional leather interiors received a new pattern design.

Two new options appeared on the options list. The GT Sport Group included the 17-inch five-spoke aluminum wheels, hood and wraparound fender stripes, a leather-wrapped shift knob (manual trans.) and an engine oil cooler.

The V-6 Sport Appearance Group, available only on the base Mustang, included 16-inch cast aluminum wheels, the rear spoiler, a leather-wrapped steering wheel and a lower bodyside accent stripe.

The 1998 interior was spruced up. A new leather pattern was used and so was the redesigned console. The clock pod on the instrument panel was removed and the clock function was integrated into the radio display. A CD player became part of the standard premium sound system as well.

Chapter 35

1999 Mustang

Production Figures
Base Coupe	73,180
Base Convertible	19,299
GT Coupe	19,634
GT Convertible	13,699
Cobra Coupe	4,040
Cobra Convertible	4,055
Total	133,637

Serial Numbers
1FALP45W6XF000001
1FA — Ford Motor Co.
L — Restraint system (L-Air bags & active belts)
P — Passenger car
45 — Body code (40-coupe, 42-GT Coupe, 44-convertible, 45-GT Convertible, 47-Cobra coupe, 46-Cobra convertible)
W — Engine code
6 — Check digit which varies
X — Model year (X-1999)
F — Assembly plant (F-Dearborn)
000001 — Consecutive unit number

Location:
Stamped on riveted plate on driver's side of dash, visible through the windshield. Certification label attached on rear face of driver's door.

Engine Codes
4 — 3.8l EFI V-6 190 hp
W — 4.6l SOHC EFI V-8 260 hp
V — 4.6l DOHC EFI V-8 320 hp (SVT Cobra)

1999 Mustang Prices*
Coupe, P40	$16,470
Convertible, P44	21,070
GT Coupe, P42	20,870
GT Convertible, P45	24,870
Cobra Coupe, P47	27,470
Cobra Convertible, P46	31,470
44U 4-speed automatic overdrive	815
54V V6 Sport Appearance Group includes: 15-in cast aluminum wheels, rear spoiler, leather-wrapped steering wheel, lower bodyside accent stripe	
Base	310
54Y 35th Anniversary Limited EditionPackage includes: 17-in 5-spoke aluminum wheels,	

black tape applique on hood,
body-color side scoops, rocker
panel moldings, rear spoiler, black
applique between taillamps,
black/silver leather/vinyl front
seats w/pony logo, silver leather
door trim inserts, silver/black floor
mats w/35th Anniv. logo, aluminum shift
knob (manual trans.)

GT	2,695
60C Convenience Group includes: Front floor mats, rear window defroster, speed control, power 6-way driver's seat	
Base/GT	550
553 All speed traction control	230
552 Brakes, anti-lock (base)	500
57Q Defroster, rear window	190
428 Emission system, high altitude	N/C
677 Mirrors, dual illum. visor	95
588 Radio, Mach 460 electronic AM/FM stereo cassette/equalizer	395
13K Rear spoiler (Std. GT)	195
63B Smoker's Package	15
64X Wheels, forged aluminum 17-in (GT only)	500
Leather seats	500

1999 Exterior Colors Codes

Chrome Yellow	BZ	**Convertible Tops**	
Rio Red	E8	Black	A
Laser Red	E9	Parchment	M
Performance Red	ES	White	W
Dark Green Satin	FU		
Atlantic Blue	K6		
Bright Atlantic Blue	K7		
Electric Green	SW		
Black	UA		
Silver	YN		
Crystal White	ZR		

1999 Interior Trim

	Medium Graphite	Oxford White	Medium Parchment	Dark Charcoal
Coupe & Convertible				
Cloth	92	-	9H	9W
Leather	T2	TZ	TH	TW
GT Coupe & Convertible				
Cloth	U2	-	UH	UW
Leather	X2	XZ	XH	XW

1999 Mustang Facts

The Mustang was extensively restyled, but the overall effect was evolutionary rather than revolutionary. Replacing the smooth rounded lines of the 1994–1998 models were creases and lines that were angular and definitive. The side sculpting was larger, leading to a taller rear side scoop. All this was done to emulate the look of the first generation Mustangs yet still maintain a contemporary '90s look. About the only thing that didn't change from the 1994–98 models was the roof. Overall, the Mustang became slightly longer and wider.

Although one can't tell by just looking at it, the rear deck lid on all 1999 Mustangs was made from sheet-molded compound to reduce weight and eliminate the possibility of corrosion.

The GT used a hood that incorporated a simulated hood scoop and also had larger, 3-inch exhaust tip extensions. The extensions measured 2.75 inches previously. All 1999 Mustangs also have a 35th Anniversary version of the tri-color emblem on the sides of the front fenders.

There were considerable refinements and improvements to the 1999 Mustang's chassis. Revised floorpan sealing and added foam in the rocker panels reduced road noise, and sub-frame connectors on the convertible reduced, as Ford put it, "mid-car shake." A 1.5-inch increase in the drive tunnel height at the rear axle resulted in more rear suspension travel.

New for the Mustang was an all-speed Traction Control System (TCS). It was designed to control wheelspin under adverse road conditions. The system uses the ABS sensors to detect when a drive wheel is spinning at a faster rate than the other wheel. To slow that wheel down, the ignition spark is retarded and the air/fuel mixture ratio adjusted. If that doesn't do the trick, the system then engages the brake on the spinning wheel, transfers power to the other drive wheel and even engages cylinder cut-off to stop the spinning condition.

In the interior, there were new fabrics and patterns and a 1-inch increase in seat travel for the driver's seat.

Although engine availability was unchanged from 1998, both Mustang engines received extensive modifications that resulted in more power. The V-6 was uprated to produce 190 hp and 5,250 rpm with 225 ft/ lbs torque at 3,000 rpm, through the use of a new intake manifold and cylinder head improvements. To improve engine smoothness, a first-order balance shaft was added.

The 4.6l modular V-8 was rated at 260 hp at 5,250 rpm with 300 ft/ lbs torque at 4,000 rpm. This was achieved through the use of higher lift and longer duration camshafts, coil-on-plug ignition, bigger valves and a revised intake manifold that increased intake flow above 2,000 rpm.

Although the five-speed manual transmission was the same T45 used in previous years, for 1999 it was manufactured under license from Borg-Warner, by Tremec. Both the V-6 and V-8 Mustangs were equipped with a 3.27:1 rear axle ratio for 1990.

A special 35th Anniversary Limited Edition Package was made available for the GT. It included 17-inch wheels, black applique trim, a black tape treatment on the hood and a special black/silver interior.

Chapter 36
2000 Mustang

Serial Numbers
1FALP45W6YF000001
1FA — Ford Motor Co.
L — Restraint system (L-Air bags & active belts)
P — Passenger car
45 — Body code (40-coupe, 42-GT coupe, 44-convertible, 45-GT convertible, 47-Cobra coupe, 46-Cobra convertible)
W — Engine code
6 — Check digit which varies
Y — Model year (Y-2000)
F — Assembly plant (F-Dearborn)
000001 — Consecutive unit number

Location:
Stamped on riveted plate on driver's side of dash, visible through the windshield. Certification label attached on rear face of driver's door.

Engine Codes
4 — 3.8l EFI V-6 190 hp
W — 4.6l SOHC EFI V-8 260 hp
V — 4.6l DOHC EFI V-8 320 hp (SVT Cobra)
H — 5.4l SOHC EFI V-8 (SVT Cobra "R")

2000 Mustang Prices

Coupe, P40	$16,520
Convertible, P44	21,370
GT Coupe, P42	21,015
GT Convertible, P45	25,270
Cobra Coupe, P47	27,605
Cobra Convertible, P46	31,605
44U 4-speed automatic overdrive	815
54V V6 Sport Appearance Group includes: 15-in cast aluminum wheels, rear spoiler, leather-wrapped steering wheel, lower bodyside accent stripe	
Base	310
60C Convenience Group includes: Front floor mats, rear window defroster, speed control, power 6-way driver's seat	
Base/GT	550
553 All speed traction control	230
552 Brakes, anti-lock (base)	500
57Q Defroster, rear window	190
428 Emission system, high altitude	N/C
677 Mirrors, dual illum. visor	95
588 Radio, Mach 460 electronic AM/FM	

stereo cassette/equalizer			395
13K Rear spoiler (Std. GT)			195
63B Smoker's Package			15
64X Wheels, forged aluminum 17-in (GT only)			500
Leather seats			500

2000 Exterior Colors — Codes

Color	Code
Sunburst Gold	BP
Laser Red	E9
Performance Red	ES
Amazon Green	SU
Atlantic Blue	K6
Bright Atlantic Blue	K7
Electric Green	SW
Black	UA
Silver	YN
Crystal White	ZR

Convertible Tops

Black	A
Parchment	M
White	W

2000 Interior Trim

	Medium Graphite	Oxford White	Medium Parchment	Dark Charcoal
Coupe & Convertible				
Cloth	92	-	9H	9W
Leather	T2	TZ	TH	TW
GT Coupe & Convertible				
Cloth	U2	-	UH	UW
Leather	X2	XZ	XH	XW

2000 Mustang Facts

Changes to the 2000 model Mustang were minimal. Three new colors, Sunburst Gold, Performance Red and Amazon Gold, were added. Three were not carried over from 1999: Chrome Yellow, Rio Red and Dark Green Satin.

Rear child seat tether anchor brackets were added to all models as was an interior deck lid release. The deck lid release had a "glow-in-the-dark" feature.

Chapter 37
1993 SVT Mustang Cobra

Production Figures
Cobra 3dr	4,993
Cobra "R" 3dr	107
Total	5,100

Serial Numbers
1FACP42D6PF000001
1FA — Ford Motor Co.
C — Restraint system (C-Air bags & active belts)
P — Passenger car
42 — Body code (42-3dr)
D — Engine code
6 — Check digit which varies
P — Model year (P-1993)
F — Assembly plant (F-Dearborn)
000001 — Consecutive Unit Number

Location:
Stamped on riveted plate on driver's side of dash, visible through the windshield. Certification label attached on rear face of driver's door.

Engine Code
D — 5.0l 235 hp

1993 SVT Mustang Cobra Prices
Cobra Hatchback	$18,505
PEP Package 250A	1,455
Mustang Cobra	
Power Equipment Group	
Manual air conditioning	
Front floor mats	
57Q Defroster, rear window	170
41H Heater, engine block immersion	20
Paint, clearcoat	91
585 Radio, AM/FM stereo/CD player	629
13C Roof, flip-open air	355
217 Seat, 4-way power driver's	183
Leather seating surfaces	523

Production By Exterior Color
Black (UA)	1,854
Vibrant Red Clearcoat (ES)	1,882

Vibrant Red (EY)		9
Teal Metallic (RD)		1,355
Total		5,100

Interior Trim Codes

JD	Black cloth
6D	Gray cloth
6C	Gray leather

Production By Interior Trim

Exterior Color	Black Cloth	Gray Cloth	Gray Leather
Black	448	327	1,079
Vibrant Red Clearcoat	361	521	1,000
Vibrant Red	1	1	7
Teal Metallic	185	414	802

1993 SVT Mustang Cobra Facts

The 1993 SVT Mustang Cobra was the first high performance model to be built by Ford's Special Vehicle Team (SVT). SVT's purpose was to produce a balanced Mustang—one that would accelerate well, and handle and brake, all wrapped up in a distinctive and exclusive package; 4,993 units were built for street use and 107 "R" models were built to specifically participate in showroom stock racing.

The 5.0l V-8 engine received numerous modifications and refinements. It was fitted with the higher-performance GT-40 cylinder heads which had larger 1.84-inch/1.54-inch intake/exhaust while the regular 5.0l heads had valves that measured 1.78 inch/1.45 inch. The larger valves were first used on the 1969 351-ci Windsor engine; in 1977, both the 302 (5.0l) and 351 engines used the small valve 302 type cylinder heads. Another major difference from the regular 5.0l engine was the use of 1.70:1 ratio Crane roller rocker arms on individual rocker studs, and in addition, the Cobra V-8 came with a stouter camshaft. And to improve reliability, the GT-40 heads came with exhaust valve seat inserts for better durability with unleaded gasoline.

The engine was rated at 235 hp at 4,600 rpm and 280 ft/lbs torque at 3,800 rpm. Valve covers used on the 1993 Cobra were fabricated from steel that was twice the production thickness.

To capitalize on the improved cylinder heads, a new upper and lower intake manifold was used. The manifold featured larger diameter runners and also used a larger 65-mm throttle body and a 70-mm mass air meter. Replacing the stock 19 lb/hr fuel injectors were 24 lb/hr units.

Taking a cue from the aftermarket industry, SVT installed smaller (by 12 percent) crank and water pump pulleys. To take advantage of all these improvements was a recalibrated Ford EEC-IV computer module. The result was 235 hp at 4,600 rpm with 280 ft/lbs torque at 4,000 rpm. The engine was redlined at 6,000 rpm.

The Cobra used the same Borg-Warner T-5 five-speed manual transmission as other production Mustangs; however, the T-5s installed in the Cobra had phosphate-coated gears and stronger bearings. In addition, the shifter was revised to produce a shorter throw and better feel. Rounding out the package was a stronger drive shaft yoke and a 3.08:1 axle ratio in the 8.8-inch limited-slip rear.

The 1993 SVT Cobra used SVT's "Controlled Compliance" suspension system. This system produced a softer ride for street use, but because Goodyear Eagle P245/45ZR17 uni-directional tires on special aluminum 17x7.5-inch wheels were used, cornering limits were still high.

Instead of using drum brakes in the rear, the Cobra used 10.07-inch vented rear disc brakes. The fronts were the same 10.84-inch vented discs as used on the GT Mustangs.

In terms of styling, the Cobra used the GT's lower grille, which housed the foglamps, but without the GT's side scoops. The front fascia featured a grille opening (the GT's was closed) that contained a running-horse emblem. In the rear was a one-piece fascia with cutouts for the exhaust tips, and a unique rear wing spoiler was mounted on the hatch/deck lid. On the rear deck lid, a Cobra emblem was affixed on the left side. The taillight bezels were the same units (modified to use the newer-style bulbs) used on the 1984–86 SVO Mustangs.

In the interior, the Cobra was the same as the Mustang GT. This included full instrumentation, articulated cloth/vinyl front sport seats with power lumbar support, premium electronic AM/FM cassette system with integrated clock and six speakers, power sideview mirrors, power windows, power locks and A/C with manual controls. What set the Cobra apart were the white-faced instruments, which have been used on all subsequent Cobras. The Cobra also came with "Cobra"-embroidered floor mats.

Few options were available with the Cobra. The listed flip-open air roof became available only on Cobras produced after February 1, 1993, and the Super Sound System, listed on the sales brochure, was never released to the public.

Each Cobra received its own signed certificate from SVT indicating its production number and VIN. This practice continued with all subsequent Cobras.

The 107 Cobra "R" models were all painted Vibrant Red Clearcoat and had the gray cloth interior. As it was primarily a Cobra built for competition use, it lacked power windows and door locks, stereo system and air conditioning. Externally, the foglamps were deleted and the Cobra wheels were painted black with a bright center cap. The "R" model's suspension was firmer than the street's version.

Chapter 38
1994 SVT Mustang Cobra

Production Figures
Cobra Coupe	5,009
Cobra Convertible	1,000
Total	6,009

Serial Numbers
1FALP42D6RF000001
1FA — Ford Motor Co.
L — Restraint system (L-Air bags & active belts)
P — Passenger car
42 — Body code (42-coupe, 45-convertible)
D — Engine code
6 — Check digit which varies
R — Model year (R-1994)
F — Assembly plant (F-Dearborn)
000001 — Consecutive Unit Number

Location:
Stamped on riveted plate on driver's side of dash, visible through the windshield. Certification label attached on rear face of driver's door.

Engine Code
D — 5.0l EFI V-8 240 hp

1994 SVT Mustang Cobra Prices*
P42 Cobra Coupe	$20,765
P45 Cobra Convertible	23,535
P45 Cobra Conv. Indy Pace Car Replica	26,845
PEP Package 250A, Coupe	1,185
Air conditioning, rear window defroster,	
front floor mats, speed control	
PEP Package 250A, Convertible	2,285
Air conditioning, rear window defroster,	
front floor mats, speed control,	
remote keyless illuminated entry,	
Mach 460 stereo system	
60K Power Group 3 (includes Power Group 2)	310
Remote keyless illuminated entry	
and cargo net	
422 emission system, Calif.	95
912 Mach 460 electronic AM/FM stereo cassette	375
917 Compact disc player (requires Mach 460	
cassette radio)	475
Leather seating surfaces (coupe)	500

*Effective 02/03/94

Production By Exterior Colors

Coupe
Black Clearcoat (UA)	1,795
Rio Red Tinted Clearcoat (E8)	2,908
Crystal White (ZF)	1,306
Total	5,009

Convertible	
Rio Red Tinted Clearcoat (E8)	1,000
Grand Total	6,009

Production By Interior Trim Codes

Exterior Color Trim Code	Black	Rio Red	Crystal White
J4 Black cloth	331	333	268
J2 Black leather	776	625	473
S2 Saddle cloth	130	208	123
S4 Saddle leather	558	742	442
SF Saddle leather (Pace Car)		1,000	

1994 SVT Mustang Cobra Facts

The 1994 Mustang was redesigned for 1994 and so was the Cobra. Only two body styles were available, a two-door coupe and convertible. The 1994 coupe had 44 percent more torsional rigidity than the comparable 1993 hatchback, while the convertible was rated at 80 percent better torsional rigidity. Both Cobra models were based on the Mustang GT.

In the front, the Cobra got its own fascia with round foglights and its own unique reflector headlamps, as opposed to the Mustang GT, which came with rectangular foglamps. Replacing the GT emblems were the Cobra snake emblems.

The Cobra also got a different rear spoiler with a built-in LED stop lamp; the stop lamp on the GT was mounted on the deck lid.

The Cobra was available in three colors: Rio Red Tinted Clearcoat, Crystal White and Black Clearcoat.

The 5.0l V-8 was essentially carried over from the 1993 Cobra, but there were some minor differences. The engine for 1994 was rated at 240 hp at 4,800 rpm, which is 5 hp more than the 1993 version. The difference was attributed to different calibration of the engine's management system. Torque output remained unchanged, but it came in at a higher rpm (4,000)

Cobra identification was used on the engine as follows: "Cobra" was cast onto the upper intake manifold, "Cobra" was embossed on the valve covers and "Cobra" was stamped on the serpentine belt and on the lower radiator hose.

The 1994 Cobra used the same upper and lower intake manifold as the 1993 model; the regular 1994 5.0l Mustang GT engine used a lower profile intake setup, which was the same one found on the Thunderbird. This was necessary because the 1994 Mustang has a lower hood. SVT chose to continue using the 1993 manifold because it produced more power; the tradeoff was the removal of the cowl-to-strut tower brace because of interference problems. The manifold featured larger diameter runners and also used a 60 mm throttle body and a 70 mm mass air meter. Replacing the stock 19 lb/hr fuel injectors were 24 lb/hr units. Bolstering the engine's durability was the addition of an engine oil cooler.

The 1994–95 Cobra came with larger tires than the 1993 model. The tires were P255/45ZR17 Goodyear Eagle GS-C's mounted on 17x8-inch cast aluminum alloy wheels. These tires were one size larger than the optional tires that could be ordered on the Mustang GT (P245/45/ZR17). The Cobra also got its own unique wheels and a 17-inch mini spare to replace the standard-size mini spare. The five-speed transmission was carried over from 1993 as well as the 8.8-inch rear axle and gear ratio.

The springs on the Cobra were softer than the GT's. The fronts were 400 lbs/in linear-rate springs while the GT's were 400–505 lbs/in variable-rate units. The rear springs were linear-rate units rated at 160 lbs/in on the Cobra and 165–265 lbs/in variable-rate units on the GT.

The front anti-sway bar on the Cobra measured 25 mm as compared to the larger 30-mm unit used on the GT; the rear bar measured 27 mm on the Cobra and 24 mm on the GT. All anti-sway bars on the 1994 Mustangs were tubular.

The Cobra was equipped with a four-wheel disc brake setup; the front discs measured 13 inches while the rears were 11.65 inches in diameter. ABS was standard on the Cobra.

In the interior, the GT's 150-mph speedometer was replaced with a 160-mph unit, and the shift knob, boot and parking brake were leather-wrapped. Cobra badging replaced the Mustang running horse on the steering wheel airbag cover. The Cobra also got its own unique floor mats.

Not so obvious were the magnesium front-seat cushion frames; these replaced the standard ones made from steel on the GT.

Options on the Cobra included a leather interior, remote keyless entry and the Mach 460 stereo/CD system. These options were all standard equipment with the 1994 Pace Car replica.

Other standard features included dual airbags, articulated sport seats with a four-way power driver's seat, premium electronic AM/FM stereo cassette and the power equipment group, which included dual electric remote control mirrors, power side windows, power door locks and a power deck lid release. Also standard was a rear window defroster, speed control, the Cobra floor mats and dual illuminated visor mirrors.

The 1994 Mustang was chosen to be the official Pace Car of the 1994 Indianapolis 500 race. Five Cobra Convertibles were built for this purpose and prepped by Jack Roush for use on the track. Three were used on the track and the other two were used for display and parade functions. The other convertibles used at the track to shuffle VIPs around were actually Mustang GTs with the Pace Car decals.

The Pace Car Cobras did not have the standard five-speed manual transmission; instead, they were fitted with Ford's four-speed AOD automatic. In addition, the cars were fitted with a roll bar, a 15-gallon fuel cell, a Halon fire extinguisher system and the usual emergency lights. The Cobra I.D. label was not used on the Pace Cars.

One thousand replicas were built for sale to the public. All were painted Rio Red and had a saddle leather interior. The convertible top was saddle as well. Besides the SVT certificate, each Pace Car got its own sequentially numbered dash emblem, located in front of the shifter.

The Pace Car decals were shipped in the trunk of each Cobra, thereby giving each owner the option of putting them on or not. One minor difference between the decals used on the Pace Cars used at the race and the replicas was the Indianapolis Motor Speedway logo. On the track cars, the wheel on the logo was white; on the replicas it was gray. Also, Cobra identification was not used on the Pace Car's valve covers.

1994 Mustang Cobra-Indy Pace Car

Chapter 39
1995 SVT Mustang Cobra

Production Figures
Cobra Coupe	4,005
Cobra Convertible	1,003
Cobra Coupe "R" Model	250
Total	5,258

Serial Numbers
1FALP42C0SF000001
1FA — Ford Motor Co.
L — Restraint system (L-Air bags & active belts)
P — Passenger car
42 — Body code (42-coupe, 45-convertible)
C — Engine code
0 — Check digit which varies
S — Model year (S-1995)
F — Assembly plant (F-Dearborn)
000001 — Consecutive unit number

Location:
Stamped on riveted plate on driver's side of dash, visible through the windshield. Certification label attached on rear face of driver's door.

Engine Codes
D — 5.0l V-8 240 hp
C — 5.8l V-8 300 hp

1995 SVT Mustang Cobra Prices**

P42 Cobra Coupe	$21,300
P45 Cobra Convertible	25,605
"R" Model	13,699
13T Convertible Hardtop	1,825
PEP Package 250A, Coupe	1,260
Air conditioning, rear window defroster	
front floor mats, speed control	
PEP Package 250C, Convertible	2,755
Air conditioning, rear window defroster,	
front floor mats, speed control, AM/FM ETR	
Mach 460 radio, leather seating surfaces	
60K Power Group 3	310
(includes Power Group 2)	
Remote keyless illuminated entry	
and cargo net	
422 Emission system, Calif.	95
912 Mach 460 electronic AM/FM stereo	375
cassette	
585 AM/FM ETR w/CD	270
Leather seating surfaces (coupe)	500
Effective 01/12/95	

Production By Exterior Color Codes

Coupe
Black Clearcoat (UA)	1,433
Rio Red Tinted Clearcoat (E8)	1,447
Crystal White (ZF)	1,125
Total	4,005

Convertible
Black (UA)	1,003

"R" Model
Crystal White (ZF)	250
Grand Total	5,258

Production By Interior Trim Codes

Coupe

Trim Code	Black	RioRed	Crystal White
J4 Black cloth	137	110	127
J2 Black leather	760	535	498
S2 Saddle cloth	55	85	66
S4 Saddle leather	481	717	434

Convertible

	Black	RioRed	Crystal White
J2 Black leather	1,003	-	-

"R" Model

	Black	RioRed	Crystal White
S2 Saddle cloth	-	-	250

1995 SVT Mustang Cobra Facts

The 1995 Cobra was virtually unchanged from the 1994 version. The convertible became a regular production model and all were painted black and had black leather interior.

A Cobra I.D. label was placed on the Cobra V-8's driver's-side valve cover.

The most interesting and unique option on the 1995 Cobra convertible was the removable hardtop. Removing and replacing the top was a two-man operation. The top came with its own carrier for placing the top when it wasn't installed on the car. The factory hardtop cannot (easily) be retrofitted to other Mustang convertibles. The "A" pillar attachment point is different (see photos) and there are rear defroster wiring differences as well. The top was originally supposed to be available on the 1994 Mustang convertible, but none were produced. It was listed as an $1,825 option. Five hundred were installed on 1995 Cobras; although it was supposed to be a Cobra-only option, several 1995 GTs were sold with the removable hardtop as well.

SVT built 250 Cobra R Mustangs for showroom stock racing in 1995. Unlike the previous 1993 R model, the 1995 R was equipped with Ford's 5.8l (351 ci) V-8 rated at 300 hp with 365 lbs/ft torque. As the Borg-Warner T5 manual transmission was not designed for so much torque, SVT used a stronger, Tremec-made unit.

The suspension on the Cobra R model featured heavier front/rear springs with larger front/rear anti-sway bars and Koni shocks and struts. The Cobra R used the same PBR brakes that are standard on the street Cobra, with larger 17x9-inch wheels mounting P255/45ZR-17 B.F. Goodrich Comp T/A tires. The Cobra R was also equipped with a larger 20-gallon fuel cell (vs. a 15.4-gallon gas tank on the stock Cobra) and a unique fiberglass hood. In keeping with its performance mission, the "R" did not have a rear seat, radio, A/C, power windows/locks/mirrors, sound insulation or fog lights. The only color available was white.

1995 Mustang Cobra

1995 Mustang Cobra "R"

Chapter 40
1996 SVT Mustang Cobra

Production Figures
Cobra Coupe	7,496
Cobra Convertible	2,510
Total	10,006

Serial Numbers
1FALP47V6TF000001
1FA — Ford Motor Co.
L — Restraint system (L-Air bags & active belts)
P — Passenger car
47 — Body code (47-coupe, 46-convertible)
V — Engine code
6 — Check digit which varies
T — Model year (T-1996)
F — Assembly plant (F-Dearborn)
000001 — Consecutive unit number

Location:
Stamped on riveted plate on driver's side of dash, visible through the windshield. Certification label attached on rear face of driver's door.

Engine Code
V — 4.6l DOHC EFI V-8 305 hp

1996 SVT Mustang Cobra Prices*
P47 Cobra Coupe	$24,810
P46 Cobra Convertible	27,580
PEP Package 250A, Coupe	1,335
Mach 460 stereo system, CD player,	
leather interior, perimeter anti-theft system	
422 Emission system, Calif.	100
534 Rear spoiler	215

*As of 03/07/96

Production By Exterior Color Codes
Coupe	
Black Clearcoat (UA)	2,122
Laser Red Tinted Clearcoat (E9)	1,940
Crystal White (ZR)	1,435
Mystic (LF)	1,999
Total	7,496
Convertible	
Black Clearcoat (UA)	1,053
Laser Red Tinted Clearcoat (E9)	962
Crystal White (ZR)	494
Total	2,509
Grand Total	10,005

Production By Interior Trim Codes

Coupe

Exterior Color	Black	Laser Red	Crystal	White Mystic
Trim Code				
J2 Black cloth	13	16	10	9
J4 Black leather	1,376	926	739	1,990
S2 Saddle cloth	8	4	10	-
S4 Saddle leather	725	994	676	-

Convertible

Exterior Color	Top Color	Interior	Quantity
Black	Black	Black Cloth	2
Black	Black	Leather	664
Black	Saddle	Leather	158
Saddle	Black	Leather	5
Saddle	Saddle	Cloth	1
Saddle	Saddle	Leather	222
White	Black	Leather	1
Laser Red	Black	Black Cloth	2
Black	Black	Leather	362
Black	Saddle	Leather	36
Saddle	Saddle	Cloth	1
Saddle	Saddle	Leather	536
White	Black	Cloth	1
White	Black	Leather	13
White	Saddle	Leather	11
Crystal White	Black	Black Leather	144
Black	Saddle	Leather	11
Saddle	Saddle	Cloth	1
Saddle	Saddle	Leather	220
White	Black	Cloth	2
White	Black	Leather	59
White	Saddle	Leather	57

Convertible Tops

Black	A
Saddle	M
White	W

1996 SVT Mustang Cobra Facts

There were some subtle changes to the Cobra's styling for 1996. Most noticeable was the domed hood, necessary to clear the new 4.6l modular engine. The taillamps were also redesigned, featuring three vertical bars, which are reminiscent of the first generation Mustang taillights. The rear spoiler was also changed, and "COBRA" lettering was stamped on the rear valance panel. Also new were 2.75-inch flared tailpipe outlets.

Because the new 4.6l modular engine was wider and taller than the 5.0l V-8, it necessitated changes to the Mustang's chassis. The No. 2 crossmember was modified to accommodate the engine's increased height and its oil pan, and the modification to the crossmember also resulted in improved structural rigidity. Also revised in this process was the front suspension's geometry. There was also a change in the power brake system for space reasons. A compact hydraulic system replaced the previous vacuum system.

The Cobra's rack & pinion system was also modified to use helical gears, and the previous plain bushings were replaced with roller type.

The Cobra continued to be equipped with 17x8-inch cast alloy wheels; but tires were changed to P245/45ZR-17 B.F. Goodrich Co. T/As. The tires were one size smaller than the Goodyear P255s used in 1995, but they were also one pound lighter each, which reduced unsprung weight.

The 1996 Cobra was equipped with the 1995 Mustang GT's higher rate springs and a thicker front anti-sway bar. The front springs were 400–505 lbs/in variable-rate units and the rear springs were also variable-rate units rated at 165–265 lbs/in. The front anti-sway bar on the Cobra measured 29 mm while the rear bar remained unchanged at 27 mm.

The biggest change was the use of Ford's 4.6l modular V-8 engine for the Cobra. Even though it had the same bore and stroke (3.55x3.54 inch) as the 4.6l modular V-8 used in the Mustang GTs, the Cobra version featured an aluminum cylinder block cast at Teksid, an Italian company located in Carmagnola, Italy. The block used six bolts to retain each nodular iron main bearing cap. Four bolts go through the top of the cap into the cylinder block; two more bolts go through the side of the block into the cap, in the same way the old FE 427-ci engines or Chrysler's 426 Hemi were cross-bolted. The engine uses a forged steel crankshaft and hot-struck powder-sintered connecting rods.

The Cobra's cylinder heads have a chain-driven, double overhead camshaft configuration with four valves per cylinder. The fuel-injection system utilizes a dual 57-mm throttle body and an 80-mm Mass Air Sensor. The 4.6l also has a built-in engine oil cooler. The water-to-oil cooler is mounted on the left side of the block and it has the oil filter mounted on its end.

The engine was assembled at Ford's Romeo, Michigan, engine plant by 12 two-person teams. Each engine has a plate affixed on the right valve cover which has the initials of the two assemblers who put that particular engine together. Output was 305 hp at 5,800 rpm with 300 ft/lbs torque at 4,800 rpm.

Along with a new engine, the 1996 Cobra got a new transmission, the Borg-Warner T45 five-speed. The transmission, which weighs in at 110 pounds, has a torque rating of 320 ft/lbs. Unlike the T5, the bell housing is an integral part of the transmission casing. This makes for a more rigid structure.

The Cobra's brakes were unchanged from 1995.

Color selection consisted of Crystal White, Black Clearcoat, Laser Red Tinted and the unusual Mystic paint. The Mystic paint exhibited four major metallic colors—green, amber, gold and purple—and the visual effect varies due to light intensity and the angle from which it is viewed.

The interior of the 1996 Cobra was a carryover from the 1995 model.

Chapter 41

1997 SVT Mustang Cobra

Production Figures

Cobra Coupe	6,961
Cobra Convertible	3,088
Total	10,049

Serial Numbers
1FALP47W6VF000001
1FA — Ford Motor Co.
L — Restraint system (L-Air bags & active belts)
P — Passenger car
47 — Body code (47-coupe, 46-convertible)
W — Engine code
6 — Check digit which varies
V — Model year (V-1997)
F — Assembly plant (F-Dearborn)
000001 — Consecutive unit number

Location:
Stamped on riveted plate on driver's side of dash, visible through the windshield. Certification label attached on rear face of driver's door.

Engine code
V — 4.6l DOHC EFI V-8 305 hp

1997 SVT Mustang Cobra Prices

P47 Cobra Coupe	$25,335
P46 Cobra Convertible	28,135
PEP Package 250A, Coupe	1,335
Mach 460 stereo system, CD player, leather interior, perimeter anti-theft system	
422 Emission system, Calif.	100
534 Rear spoiler	195

Production By Exterior Color

Coupe	
Black Clearcoat (UA)	2,369
Rio Red Tinted Clearcoat (E8)	1,994
Crystal White (ZR)	1,543
Pacific Green Clearcoat (PS)	1,055
Total	6,961
Convertible	
Black Clearcoat (UA)	1,180
Rio Red Tinted Clearcoat (E8)	925
Crystal White (ZR)	606
Pacific Green Clearcoat (PS)	377
Total	3,088
Grand Total	10,069

Production By Interior Trim Codes

Coupe

	Black	Rio Red	Crystal White	Pacific Green
J7 Black cloth	43	39	20	-
J4 Black leather	1,641	1,131	817	-
S7 Saddle cloth	7	13	47	27
S4 Saddle leather	678	811	659	1,028

Convertible

Exterior Color	Top Color	Interior	Quantity
Black	Black	Black Cloth	14
Black	Black	Leather	840
Black	Saddle	Cloth	2
Black	Saddle	Leather	92
Saddle	Black	Leather	12
Saddle	Saddle	Leather	219
White	Black	Leather	1
Rio Red	Black	Black Cloth	6
Black	Black	Leather	416
Black	Saddle	Leather	15
Saddle	Saddle	Cloth	7
Saddle	Saddle	Leather	466
White	Black	Leather	13
White	Saddle	Leather	2
Crystal White	Black	Black Cloth	2
Black	Black	Leather	190
Black	Saddle	Cloth	1
Black	Saddle	Leather	4
Saddle	Black	Leather	7
Saddle	Saddle	Cloth	4
Saddle	Saddle	Leather	258
White	Black	Leather	76
White	Saddle	Cloth	3
White	Saddle	Leather	61
Pacific Green	Saddle	Saddle Cloth	4
Saddle	Saddle	Leather	365
White	Saddle	Leather	8

Convertible Tops

Black	A
Saddle	M
White	W

1997 SVT Mustang Cobra Facts

The 1997 Cobra was basically unchanged, save for a slight change in the front upper grille opening. This opening ducted more air to the new cross-the-board Mustang cooling system. All models had a wider and taller radiator and a larger diameter fan. Exclusive to the Cobra was a new parallel-flow air conditioning condenser.

Color choice for the 1997 consisted of Crystal White, Black Clearcoat, Rio Red Tinted Clearcoat and a new color, Pacific Green Clearcoat Metallic.

Chapter 42
1998 SVT Mustang Cobra

Production Figures
Cobra Coupe	5,174
Cobra Convertible	3,480
Total	8,654

Serial Numbers
1FALP47V6WF000001
1FA — Ford Motor Co.
L — Restraint system (Air bags & active belts)
P — Passenger car
47 — Body code (47-coupe, 46-convertible)
V — Engine code
6 — Check digit which varies
W — Model year (W-1998)
F — Assembly plant (F-Dearborn)
000001 — Consecutive unit number

Location:
Stamped on riveted plate on driver's side of dash, visible through the windshield. Certification label attached on rear face of driver's door.

Engine Code
V— 4.6l DOHC EFI V-8 305 hp

1998 SVT Mustang Cobra Prices*
Cobra Coupe, P47	$25,710
Cobra Convertible, P46	28,510
54E Electronic Leather/Trim Group	
Anti-theft system, Mach 460 ETR w/cassette,	
leather seating surfaces	
Coupe	NC
Conv	1040
18A Anti-theft system (requires 54E)	145
422 Emission system, Calif.	170
534 Rear spoiler	195

* as of March 15, 1998

Production By Exterior Color

	Coupe	Convertible
Black Clearcoat (UA)	1,708	1,256
Laser Red Tinted Clearcoat (E9)	1,236	842
Crystal White (ZR)	958	578
Atlantic Blue Clearcoat (PS)	563	249
Canary Yellow Clearcoat (BZ)	709	555
Total	5,174	3,480
Grand Total		8,654

Production By Interior Trim Codes

Coupe	Black	Laser Red	Crystal White	Atlantic Blue	Canary Yellow
J7 Black cloth	6	5	8	10	5
J4 Black leather	1,344	675	506	553	704
S7 Saddle cloth	2	5	10	-	-
S4 Saddle leather	356	551	434	-	-

Convertible Exterior Color	Top Color	Interior	Quantity
Black	Black	Black Leather	967
Black	Saddle	Leather	61
Saddle	Black	Leather	11
Saddle	Saddle	Leather	215
White	Black	Cloth	1
White	Black	Leather	1
Laser Red	Black	Black Cloth	1
Black	Black	Leather	340
Black	Saddle	Leather	12
Saddle	Saddle	Cloth	2
Saddle	Saddle	Leather	484
White	Black	Leather	3
Crystal White	Black	Black Leather	218
Black	Saddle	Leather	6
Black	Saddle	Cloth	1
Saddle	Saddle	Leather	267
White	Black	Leather	46
White	Saddle	Cloth	1
White	Saddle	Leather	40
Atlantic Blue	Black	Black Cloth	1
Black	Black	Leather	224
White	Black	Cloth	1
White	Black	Leather	23
Canary Yellow	Black	Black Cloth	5
Black	Black	Leather	542
White	Black	Leather	8

1998 SVT Mustang Cobra Facts

The changes on the 1998 Cobra can be described as minimal, at best. Visually, the wheels were the most noticeable change these were the same as those used on the 1995 "R" model Cobra, except that the wheel cutouts were painted gray.

The 1998 interior was spruced up. A new leather pattern was used as well as a redesigned console. The clock pod on the instrument panel was removed, and the clock function was integrated into the radio display. A CD player became part of the standard premium sound system as well.

Color choice for the 1998 consisted of Crystal White, Black Clearcoat, Laser Red Tinted Clearcoat and two new colors, Atlantic Blue Clearcoat and Canary Yellow Clearcoat.

Chapter 43

1999 SVT Mustang Cobra

Production Figures
Cobra Coupe	4,040
Cobra Convertible	4,055
Total	8,095

Serial Numbers
1FALP47V6XF000001
1FA — Ford Motor Co.
L — Restraint system (L-Air bags & active belts)
P — Passenger car
47 — Body code (47-coupe, 46-convertible)
V — Engine code
6 — Check digit which varies
X — Model year (X-1999)
F — Assembly plant (F-Dearborn)
000001 — Consecutive unit number

Engine Code
V— 4.6l DOHC EFI V-8 320 hp

Location:
Stamped on riveted plate on driver's side of dash, visible through the windshield. Certification label attached on rear face of driver's door.

1999 SVT Mustang Prices
Cobra Coupe, P47	$27,470
Cobra Convertible, P46	31,470
13K Rear spoiler	195
63B Smoker's Package	15

Production By Exterior Color

Coupe
Black (UA)	1,619
Laser Red (E9)	1,292
Ultra White Clearcoat (ZR)	794
Electric Green Clearcoat Metallic (SW)	408
Total	4,113

Convertible
Ebony Clearcoat (UA)	1,755
Rio Red Tinted Clearcoat (E9)	1,251
Ultra White Clearcoat (ZR)	731
Electric Green Clearcoat Metallic (SW)	318
Total	4,055
Grand Total	8,168

Production By Interior Trim Codes
Coupe

Interior	Black	Laser Red	Crystal White	Electric Green
DW Dark Charcoal leather	1,204	734	472	185
DH Medium Parchment leather	412	458	332	223
Other	3	7	-	10

Convertible Exterior Color	Top Color	Interior	Quantity
Black	Black	Dk. Charcoal	1,287
Saddle		Dk. Charcoal	9
White		Dk. Charcoal	8
Black		Parchment	134
Saddle		Parchment	317
Rio Red	Black	Dk. Charcoal	598
White		Dk. Charcoal	16
Black		Parchment	28
Saddle		Parchment	608
Other		Other	1
Ultra White	Black	Dk. Charcoal	278
White		Dk. Charcoal	63
Black		Parchment	2
Saddle		Parchment	349
White		Parchment	38
Other		Other	1
Electric Green	Black	Dk. Charcoal	89
Saddle		Dk. Charcoal	9
White		Dk. Charcoal	97
Black		Parchment	175
Saddle		Parchment	1,482
White		Parchment	38

1999 SVT Mustang Cobra Facts

Besides the 1999 styling changes common to all Mustangs, the Cobra came with round foglights in the lower front fascia and a regular hood, whereas the GT used a hood that incorporated a simulated hood scoop. Both the Cobra and GT had 3-inch exhaust tip extensions. All 1999 Mustang Coupes got an SMC (Sheet Molded Compound) rear deck lid to save some weight.

Another difference between the GT and the Cobra was that the grille running-horse emblem wasn't surrounded by a chrome band. Cobras also do not have the 35th Anniversary tri-color bar emblem on the sides of the front fenders.

In the interior, there was new upholstery patterns and colors. The 1999 Cobra interior color choices were Dark Charcoal and Medium Parchment—both in leather. There were no cloth interiors available in 1999. Although not exclusive to the Cobra, but important nonetheless, was the 1-inch-longer track for the driver's seat which was helpful for taller drivers.

Exterior colors were Ultra White Clearcoat, Ebony Clearcoat, Rio Red Tinted Clearcoat and Electric Green Clearcoat Metallic.

The 4.6l DOHC modular V-8 remained as the Cobra's sole powerplant. The engine, however, put out 15 hp and 17 ft/lbs torque over the 1998 version. This was due to different intake port geometry and a redesigned combustion chamber. These changes were said to make the fuel mixture "tumble" into the combustion chamber, thereby promoting better combustion. In addition, the engine benefited from new, stronger main and rod bearings.

The 1999 ignition system was changed to a coil-on-plug system—again to aid the combustion process. A better type of knock sensor, a differential linear type, replaced the former resonant knock sensor to better control any impending detonation.

Although the Cobra was supposed to be making 15 hp more, several enthusiast magazines reported that the 1999 Cobra seemed slower than previous models. Several Cobra owners had their cars dyno tested and the engine was down on power.

On August 6, 1999, Ford stopped the sale of any unsold Cobras sitting on dealers' lots and recalled the rest to replace the intake manifold, the engine management computer and the entire exhaust system from the catalytic converter back.

There were major changes in the Cobra's drivetrain. The transmission and rear axle and ratio were carryovers from the 1998 Cobra; the differential case, though, was aluminum. The transmission was the same T45 five-speed manual as on 1998 models, but it was manufactured by Tremec, under license from Borg-Warner for 1999. Tremec-built T45s have the word "Tremec" cast into the housing.

New for the Mustang and standard equipment on the Cobra was an all-speed Traction Control System (TCS). The system was designed to control wheelspin under adverse road conditions.

There were considerable refinements and improvements to the 1999 Mustang's chassis. Revised floorpan sealing and added foam in the rocker panels reduced road noise, and sub-frame connectors on the convertible reduced "mid-car shake." A 1.5-inch increase in the drive tunnel height at the rear axle resulted in more rear suspension travel.

The most significant change on the Cobra was the use of a new independent rear suspension system. The system used short and long arms that mount on a tubular subframe. The lower arms are aluminum while the upper arms are steel. The subframe also holds the aluminum differential case, which was borrowed from the Lincoln Mark VIII.

The independent-rear mounts at the very same four mounting points as the regular Mustang solid-axle suspension. Although the whole IRS setup weighs 80 pounds more than a comparable straight axle rear, there is a 125-pound decrease in unsprung weight with the IRS. This results in better ride and handling.

The independent rear design allowed for much stiffer rear springs and the Cobra uses 470 lbs/in springs along with a 26-mm tubular anti-sway bar. The Mustang GT uses 210 lbs/in springs with a 23-mm solid bar.

The front suspension configuration remained unchanged, but the front strut's position was changed. The front spring rate was increased to 500 lbs/in (vs. 450 lbs/in on the base and GT). Front sway bar diameter was 28 mm.

The Cobra's steering system was also refined. The gear ratio was increased to 15.0:1 while the number of turns, lock to lock, were reduced to 2.5. The turning circle was reduced to 37.9 feet.

The Cobra's brakes were unchanged. The Cobra's front vented discs measured 13 inches, while the rears were 11.65 inches in diameter. ABS again was standard on the Cobra.

Chapter 44
2000 SVT Mustang Cobra

Serial Numbers
1FALP47V6YF000001
1FA — Ford Motor Co.
L — Restraint system (L-Air bags & active belts)
P — Passenger car
47 — Body code (47-coupe, 46-convertible)
V — Engine code
6 — Check digit which varies
Y — Model year (Y-2000)
F — Assembly plant (F-Dearborn)
000001 — Consecutive unit number

Location:
Stamped on riveted plate on driver's side of dash, visible through the windshield. Certification label attached on rear face of driver's door.

Engine Codes
V — 4.6l DOHC EFI V-8 320 hp
H — 5.4l SOHC EFI V-8 385 hp

2000 SVT Mustang Cobra Prices
Cobra Coupe, P47	$27,605
Cobra Convertible, P46	31,605
428 Emission system, high altitude	N/C
13K Rear spoiler	195

Exterior Colors / Codes
Exterior Colors	Codes
Silver	YN
Atlantic Blue	K6
Laser Red	E9
Black	UA
Crystal White	ZR

Convertible Tops
Black	A
Parchment	M
White	W

2000 Interior Trim
Medium Dark
Parchment Charcoal

Coupe & Convertible
Leather DH DW

2000 SVT Mustang Cobra Facts
As of this writing, a 2000 "R" model Cobra is scheduled for release

by the spring of 2000. Following the same pattern as the 1993 and 1995 "R" model versions, the 2000 model was not equipped with any luxury options such as air conditioning, stereo, power windows, and power door locks. The standard seats were replaced with special Recaro versions which also incorporated a Cobra snake and "R" emblem on the headrest area. The snake/"R" emblem was also used on the reverse indicator letter on the transmission shifter.

The engine used on the Cobra was a specially modified version of Ford's 5.4l SOHC V-8, rated at 385 hp, and mated to a T45 six-speed manual transmission. All "R" models were painted red and production was limited to 300 units.

Visually, the "R" model was distinguished by the tall rear wing spoiler and the front air dam. The low front air dam, which limited ground clearance, was easily removeable for street use.

2000 Mustang Cobra "R"

Chapter 45

1965 Shelby Mustang

Production Figures

Street prototype	1	Competition models (R)	36
Street production models	516	Drag cars	9
Competition prototypes	2	Total	561

Shelby VIN
SFM5S001
SFM — Shelby Ford Mustang
5 — Year (1965)

S — Street, R — Race
001 — Consecutive unit number, 001 to 562

Location
On pop-riveted plate on driver's side inner fender panel over Ford VIN; also stamped on passenger's side inner fender panel halfway between firewall and radiator.

Ford VIN
5R09K000001
5 — Last digit of model year
R — Assembly plant
(R-San Jose)

09 — Body code (09-2dr fastback)
K — Engine code
(271 hp 289 ci V-8)
000001 — Consecutive unit number

Location
Underneath Shelby VIN plate; on driver's and passenger's side inner fender panels at outside edge near shock tower; can only be viewed if fenders are removed; on original engine block, beneath front exhaust port on passenger side.

Carburetor
Holley R-3259

Distributor
C5GF-12127-A, C50F-12127-E, C5AZ-12127-EEZ

1965 Shelby Prices

	Retail
Street	$4,547.00
Race	5,995.00

1965 Shelby Colors
All 1965 Shelby Mustangs were painted Wimbledon White, code M. Interior was black.

1965 Shelby GT350 Facts
1965 Shelby Mustangs can be identified by their blue rocker panel GT350 stripes. The wide blue racing stripes running the length of the car were mostly dealer installed but some were installed at the factory. With the exception of the stock Mustang

gas cap and relocated grille emblem (the far left of the grille), all other Mustang emblems were removed. Some GT350s did have a GT350 emblem on the left side of the taillight panel. The Shelby's hood was different from stock as it was made from fiberglass and has a functional hood scoop. Some hoods came with steel frames and fiberglass skins.

Most noticeable in the interior was the pod attached to the dash top which housed a tachometer and oil pressure gauge. All 1965s did not have a rear seat; a fiberglass shelf took its place and was also the location for the spare tire. Large, 3 inch competition seatbelts replaced the stock belts. Three types of wooden steering wheels were used. Early cars came with 16 inch diameter wheels with slotted spokes; later cars came with 15 inch wheels with either slotted spokes or three holes in each spoke.

The engine was the stock High Performance 271 hp 289 with Shelby modifications which boosted horsepower to 306. These were an aluminum high-rise intake manifold with a 715 cfm Holley four-barrel carburetor and steel Tri-Y headers exiting to a dual exhaust system using glasspack mufflers and exiting in front of the rear wheels. The exception here was cars built after July 6, 1965 for delivery to California, Florida or New Jersey which got a rear-exiting exhaust system. Cobra valve covers and a larger capacity aluminum Cobra oil pan rounded out the package.

Transmission was an aluminum case Borg Warner T-10. The 9 inch rear had 3.89 gears (but others were available) and they all had a Detroit Locker differential. A drive shaft safety loop was standard equipment. Additional rear axle control was achieved through the use of traction bars and travel-limiting cables.

Shock absorbers were the renowned Koni adjustable units. The use of a Monte Carlo bar, attached between the shock towers and a one-piece export brace, noticeably stiffened the GT350's front frame structure. All cars came with a lowered front suspension (1 inch) and, for better weight distribution, the battery was relocated to the trunk, at least for most cars between numbers 001 and 324.

Manual brakes and manual quick ratio steering were standard. Stock wheels were silver painted steel rims with chrome lug nuts with Goodyear 7.75x15 Blue Dot tires. Optional were the Cragar mags.

The "R" (race) version, in addition to all the street version features, got a fiberglass front lower apron, engine oil cooler, larger capacity radiator, front and rear brake cooling assemblies, 34 gallon gas tank, 3½ inch quick fill gas cap, electric fuel pump, large diameter exhaust pipes with no mufflers, five magnesium 15x7 wheels, revised wheel openings, Interior Safety Group (roll bar, shoulder harness, fire extinguisher, flame resistant interior, plastic rear window, aluminum framed sliding plastic side windows) complete instrumentation (tachometer, speedometer, oil pressure and temperature, water temperature and fuel pressure), and final track test and adjustments. The engine was further modified to produce 350 hp.

Cars with serial numbers 004-034 have no "S" designation on the VIN plate. All others are either "S" or "R."

Chapter 46

1966 Shelby Mustang

Production Figures
GT350	1,368
GT350H Hertz	1,001
GT350 Drag cars	4
GT350 Convertibles	4*
Total	2,378

*Does not include the 12 continuation convertibles built during 1980 through 1982

Shelby VIN
SFM6S0001
SFM — Shelby Ford Mustang
6 — Last digit of model year
S — Street car
0001 — Consecutive unit number, 0001 to 2380

Location
On plate pop-riveted on driver's side inner fender panel covering Ford VIN; also on passenger's side inner front fender panel halfway between radiator and firewall.

Ford VIN
6R09K000001
6 — Last digit of model year
R — Assembly plant (R-San Jose)
09 — Body code (09 - Mustang 2dr 2+2)
K — Engine code, (271 hp 289 ci)
000001 — Consecutive unit number

Location
Underneath Shelby VIN plate; stamped on driver's and passenger's side inner fender panel, near shock tower at outside edge; can only be seen with fender removed; stamped on engine block, beneath front exhaust port on passenger side.

Carburetors
Holley R-3259
Ford C30F-9510-AB, AJ, C40F-9510-AD, AL, AT, or C4ZF-9510-G

Distributors
C5GF-12127-A, C50F-12127-E, C5AZ-12127-EEZ

1966 Shelby Prices

	Retail
Shelby GT350	$4,428.00
High performance Ford automatic transmission	N/C
Fold-down rear seat	50.00
AM radio	57.50
Alloy wheels	268.00*
Stripe	62.50
Detroit No-Spin rear axle unit	141.00
Cobra Supercharger (Paxton)	670.00

*Approximate price

1966 Shelby Exterior Colors
Wimbledon White
Candyapple Red
Sapphire Blue
Ivy Green
Raven Black

1966 Shelby Interior Trim
Black

1966 Shelby GT350 Facts

Visually, functional rear quarter panel scoops were added. The stock Mustang air extractor louvers were replaced with windows. All white GT350s came with blue rocker panel stripes. Other colors came with white rocker panel stripes. GT350H cars came with gold stripes with the exception of some early white Hertz cars which got blue stripes.

Some cars came with all steel hoods replacing the fiberglass/steel support hoods. A GT350 gas cap replaced the stock Mustang cap.

As with the 1965s, most of the Le Mans stripes were dealer installed, matching the rocker panel stripes in color.

In the interior, because the standard Mustang five-dial gauge panel was used, the center dash pod was eliminated. In its place stood a 9000 rpm Cobra tachometer. The Mustang rear seat replaced the fiberglass shelf, and with the exception of just 82 cars, all 1966 GT350s had the fold-down rear seats. The standard steering wheel was the Mustang optional Deluxe wheel with a GT350 logo.

Only the first 252 1966 cars had the lowered "A" arms because these were actually leftover 1965s updated for 1966. As a cost-cutting measure, subsequent cars used the standard mounting points. For the same reasons, the override traction bars were replaced on cars numbering 800 and above with bars that mounted beneath the axle, but as with most Shelbys, there were exceptions.

The rear exiting exhaust system replaced the side pipes of 1965.

GT350s with the three-speed automatic transmission came with a Ford 600 cfm carburetor replacing the Holley 715.

Relegated to the option list, factory or dealer, were Koni shocks, the Detroit Locker differential, and the wood-rimmed steering wheel.

Wheel selection was greater for 1966. The leftover cars came with either the silver-painted steel wheels or the Cragar mags, which could also be had on later cars. The standard wheel on later cars was a silver-painted 14 inch Magnum 500, but the Hertz cars came with chrome versions. The 14 inch Shelby aluminum ten-spoke wheels became optional on later cars. Some cars also came with plain 14 inch silver-painted wheels.

Early cars generally came with hollow-letter Cobra valve covers while later cars had solid letters and black crinkle finish.

One thousand GT350s were sold to Hertz and designated GT350H. Most were painted black and most had gold stripes. Some were equipped with the four-speed manual but the great majority came with the automatic transmission. Chrome Magnum 500 wheels were standard on these cars.

In April 1966, the Paxton Supercharger was made available as an option.

Six Shelby convertibles were built but not available for sale to the public. All had the automatic transmission and air conditioning. The side scoops were not functional as they would have interfered with the convertible top mechanism. Twelve continuation convertibles were built during 1980-82 to cash in on the Shelby phenomenon. The Shelby organization was still a functioning entity so these cars, built on refurbished original "K" convertibles, got the same Shelby identification as the original six.

1966 Shelby GT350H

GT500KR Fastback	4,472.57
GT500KR Convertible	4,594.09
Power disc brakes	64.77
Power steering	84.47
Shoulder harness	50.76
Fold-down rear seat (Fastbacks only)	64.78
Radio, AM push-button	57.59
Select-O-Matic transmission	50.08
Tinted glass (air conditioned cars only—required)	30.25
Tilt-away steering wheel	62.18

1968 Shelby Exterior Colors

Color	Code
Raven Black	A
Lime Green Metallic	I
Wimbledon White	M
Medium Blue Metallic	Q
Dark Green Metallic	R
Candyapple Red	T
Meadowlark Yellow	W
Dark Blue Metallic	X
Gold Metallic	Y
Orange	

1968 Shelby Interior Trim

Trim	Code
Black	6A
Saddle	6F

Convertible Top Colors
Black
White

1968 Shelby Facts

Production of the Shelby continued, but under Ford control and at the A.O. Smith Company facility in Livonia, Michigan. A new body style joined the line-up—the convertible available in either a GT350 or GT500. Officially, the 1968 cars were renamed Shelby Mustang Cobra GT350/GT500/GT500KR, reflecting Ford's proclivity to use the Cobra name in all its performance applications.

Although there wasn't much of a difference mechanically between a 1967 and a 1968 Shelby, the 1968 nose was restyled. The new look was decidedly Mustang, yet with a much more aggressive look. Fiberglass was again used to create the new front end treatment. The grille opening housed either Lucas or Marchal foglamps while the hood used twin front hood scoops with a set of rear hood louvers.

New Cobra emblems were used on the front fenders and on the passenger side of the dash panel. The interior was Deluxe Mustang in either black or saddle but the Shelby used a console which housed two Stewart Warner gauges—oil pressure and amps. The console storage compartment had a Cobra embossed top. The roll bars with the inertia-reel harnesses were used on all Shelbys with the exception of the convertible, which used a unique padded bar.

Standard wheels were steel with a mag style wheel cover. Optional were the ten-spoke Shelby wheels. These wheels were cast differently than the 1967 versions and will have ball joint interference if installed on a 1967 Shelby.

The GT350 lost some of its zip as the High Performance 289 was replaced by a production 302 V-8. The 302 did use an aluminum Cobra intake manifold and Holley 600 cfm carburetor for a 250 hp rating. The 302 used a Cobra oval air cleaner and Cobra valve covers. Functional Ram Air was optional. The Paxton Supercharger was optional on the GT350; additional gauges—fuel pressure and boost—were mounted on the console.

The GT500 again got the 428 ci Police Interceptor V-8 rated at 360 hp even though it came with only a single 715 cfm Holley carburetor mounted on an aluminum manifold. A few GT500s have the 427 Low Riser engine which was available only with an automatic transmission. The letter "W" must be present in the VIN for an original factory installed 427.

The GT500KR replaced the GT500 when the 428 Cobra Jet engine became available. Underrated at 335 hp, it put out close to 400 hp and the GT500KR had the same features found on the 1968½ Cobra Jet Mustang. KRs got Cobra Jet emblems on the fenders, dash and gas cap lid. All KRs had functional Ram Air which meant no Cobra oval aircleaner.

1968 Shelby GT500 convertible

1968 Shelby GT500KR Fastback

Chapter 49

1969-70 Shelby Mustang

Production Figures*

Barrier Test & Prototype Pilot cars	3
GT350 Fastbacks	935
GT500 Fastback Hertz cars	150
GT350 Convertibles	194
GT500 Fastback	1,536
GT500 Convertibles	335
Total—1969–1970 models	3,153*

*789 are updated 1970 models

Serial Numbers
9F02M480001
9 — Last digit of model year (0-1970 updated cars)
F — Assembly plant (F-Dearborn)
M — Engine code (M-351, R-428CJ-R)
48 — Shelby code
0001 — Consecutive unit number

Location
Plate riveted to dash panel on driver's side, visible through windshield; stamped on warranty plate located on face of driver's door, which also reads "Special Performance Vehicle"; stamped on driver's and passenger's inner fender panel halfway between shock tower and firewall, visible with fenders removed; additional plate stating "Custom-Crafted by Shelby Automotive, Inc." attached above warranty plate.

Distributors
GT350- C90F-12127-M or N/manual, C90F-12127-M or T/automatic
GT500- C8AF-12127-T

Carburetors
GT350- C9ZF-9510-C/manual, C9ZF-9510-D/automatic
GT500- Holley R-4279/manual, R-4280/automatic

1969 Shelby Prices

	Retail
GT350 SportsRoof	$4,434.00
GT350 Convertible	4,753.00
GT500 SportsRoof	4,709.00
GT500 Convertible	5,027.00

Close ratio 4-speed transmission (GT350 only, std. GT500)	N/C
Automatic transmission	30.54
Air conditioning	374.39
Optional axle ratio	6.13
Traction-Lok differential	60.97
Drag Package	155.45
Sport Deck rear seat (fastback only)	91.51
Tilt-away steering wheel	62.24
Power ventilation	37.83
AM radio	57.38
AM/FM stereo	170.76
Stereo tape (requires AM radio)	125.64
Intermittent windshield wipers	15.85
Tinted glass	30.54
Heavy-duty batteries (GT350 only)	
Option #1	7.93
Option #2	15.85
F60x15 Goodyear tires, extra heavy-duty Suspension Package	60.97

1969-70 Shelby Exterior Colors

Color	Code
Acapulco Blue	D
Black Jade	C
Silver Jade	4
Gulfstream Aqua	F
Pastel Gray	6
Candyapple Red	T
Royal Maroon	B
Grabber Blue	J
Grabber Green	Z
Grabber Yellow	
Grabber Orange	U

1969-70 Shelby Interior Trim

Trim	Code
Black	3A
White	3W
Red	3D

Convertible Top Colors

Black
White

1969-70 Shelby Facts

This was the last year for the Shelby Mustangs. Based on the new SportsRoof and convertible Mustang body styles, the Shelby's styling was unique and bore little resemblance to the production Mustang. Two models were available, the GT350 and GT500, each in a fastback or convertible.

The front end styling was a complete departure from previous Shelby Mustangs. Fiberglass fenders and hood created a large rectangular grille opening which houses two 7 inch headlights. Lucas foglamps were mounted beneath the bumper. The hood had three forward-facing NASA scoops (the center one providing air to the engine's intake system) and two rear-facing scoops. The front fenders also had brake scoops, as did the rear, which provided air to the brakes. The convertibles used a rear scoop that was mounted lower to prevent interference with the convertible top mechanism. The rear of the car used a fiberglass deck lid and extensions to form a pronounced spoiler. 1965 Thunderbird lights were utilized, and the Shelby used a unique aluminum exhaust collector exiting in the center beneath the bumper.

Side stripes ran the length of the car's sides with either GT350 or GT500 lettering at the front fender in front of the brake scoop. Snake emblems were located behind the rear side windows and on the left side of the front grille. Cobra Jet emblems, like the ones on the 1968 GT500KR, were used on the GT500's front fenders.

The interior was the Mustang's Deluxe Interior Decor Group, in black or white, with Shelby identification on the door panels, steering wheel and passenger's dash. The console top had two Stewart Warner gauges, oil pressure and amps, and two toggle switches for the foglamps and courtesy lights. The instrument cluster contained temperature, 8000 rpm tachometer, 140 mph speedometer and fuel gauges. All fastback Shelbys have roll bars with the inertia-reel harnesses while the convertibles used the same 1968 type roll bar. A small number of Shelby Mustangs came with red interiors.

The GT350 came with the 290 hp four-barrel version of the 351 Windsor engine. The only difference between it and other 351s was the use of an aluminum intake manifold and Cobra valve covers.

The GT500 used the 428CJ-R V-8 with all the regular Mustang variations applying to the Shelby (see 1969 Mustang).

The Competition Suspension, transmission and rear axle options paralleled those found on the Mach 1.

Wheels were unique to the Shelby. The 15x7 inch rims used an aluminum center section welded to a chrome steel rim. Standard tires were E70x15; most cars came with F60x15 Goodyear Polyglas GTs.

Other standard features included power steering, power front disc brakes, and four-speed manual transmission.

About 789 cars were unsold in 1969. These were "updated" and sold as 1970 models by changing the VIN to reflect 1970 as the model year. Other changes were the twin black hood stripes and a Boss 302 type chin spoiler.

1969 Shelby GT500 convertible

Appendix

Warranty Plates and Certification Labels
Three different Warranty Plates were used on 1965-69 Mustangs; however, they all displayed the same information. The warranty number was the same as the Mustang's VIN, while the rest of the codes indicated body style, color code, trim code, date built code, District Sales Office (DSO), axle and transmission codes. If the color code was missing, that meant that this Mustang came with a Special Order non-stock color.

From 1970-78, a label replaced the metal plate with the only major difference being the month and year were shown rather than date and month for the build date.

During 1979-80, the label was again revised to include additional information such as air conditioning code, vinyl roof color code and weight. In addition to month and year, scheduled date was included showing date and month of assembly.

The label was again changed slightly from 1981-90 to include suspension, sunroof/moonroof, and bodyside molding codes.

Casting Dates and Manufacturing Dates
All Mustang parts have date codes that are either cast or stamped or both on that part. Date codes that are cast into the part indicate the date the part was cast. Manufacture dates are stamped into the part. These date codes are fairly simple to decipher. For example, if a part has the date 9 B 16, it can be understood as follows:

9 — year, in this case, 1969
B — month, February
16 — day of month

Parts installed into a Mustang were cast or manufactured before the car was actually built. This means that date codes could be as much as thirty days before actual vehicle manufacture. This may be important in ascertaining a Mustang's originality.

Sheet Metal Date Codes
In much the same way, Mustang sheet metal was stamped to show the date it was manufactured. For example, a sheet metal part with 10 5 D 1 breaks down as follows:

10 — Month(October)
5 — Date
D — Stamping plant(Dearborn)
1 — Shift (first shift)

Although sheet metal date codes do not indicate year of manufacture, as with other components, the stamping date will fall before vehicle manufacture, up to 30 days.

1965

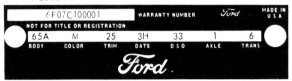

1966–68

1969

```
MANUFACTURED BY                         100001
FORD MOTOR COMPANY

    08/70 THIS VEHICLE CONFORMS
    TO ALL APPLICABLE FEDERAL
    MOTOR VEHICLE SAFETY STAN-
    DARDS IN EFFECT ON DATE OF
    MANUFACTURE SHOWN ABOVE
```

VEH. IDENT NO.		BODY	COL.
1F01M100001		65D	P
TRIM	**AXLE**	**TRNS.**	**DSO**
1R	6	U	11

NOT FOR TITLE OR REGISTRATION

MADE IN U.S.A.

1970–78

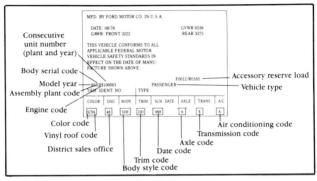

1979–80

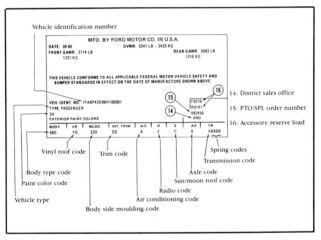

1981–90

Third generation Mustang sheet metal is dated the same way. However, replacement sheet metal will also have a month and year code, such as 9 85 (September 1985), preceding the regular date code.

District Codes (DSOs)

Units built on a Domestic Special Order, Foreign Special Order or other special orders will have the complete order number in this space. Also to appear in this space is the two-digit code number of the district which ordered the unit. If the unit is a regular production unit, only the district code number will appear.

That is the typical DSO explanation found in Ford shop manuals. A DSO is an internal Ford production code which groups batches of similarly equipped cars to be built at the same time so that the specific parts used would be correctly scheduled to be on the assembly line as this group of cars was rolling down the line. Thus a group of Mustang LXs destined for the California Highway Patrol would get a specific DSO number plus the two-digit code for L.A. (for example) which is 71.

Code	District	Code	District	Code	District
11	Boston	41	Chicago	71	Los Angeles
13	New York	43	Milwaukee	72	San Jose
15	Newark	44	Twin Cities	73	Salt Lake City
16	Philadelphia	45	Davenport	74	Seattle
17	Washington	46	Indianapolis	75	Phoenix
21	Atlanta	47	Cincinnati	81	Ford of Canada
22	Charlotte	51	Denver		
24	Jacksonville	52	Des Moines	83	Government
25	Richmond	53	Kansas City	84	Home Office Reserve
27	Cincinnati	54	Omaha		
28	Louisville	55	St. Louis	85	American Red Cross
32	Cleveland	56	Davenport		
33	Detroit	61	Dallas	87	Body Company
34	Indianapolis	62	Houston		
35	Lansing	63	Memphis	89	Transport Services
37	Buffalo	64	New Orleans		
38	Pittsburgh	65	Oklahoma City	90-99	Export

Date Codes

Month	Code first year	Code second year (if model year exceeds twelve months)
January	A	N
February	B	P
March	C	Q
April	D	R
May	E	S
June	F	T
July	G	U
August	H	V
September	J	W
October	K	X
November	L	Y
December	M	Z

Mustang transmissions 1965-73

	Code
Three-speed manual	1
Four-speed manual, wide ratio	5
Four-speed manual, close ratio	6 (1967-71)
Four-speed manual, close ratio	E (1972-73)

Mustang transmissions 1965-73

	Code
Three-speed automatic *C4*	W
Three-speed automatic *FMX*	X
Three-speed automatic *C6*	U

Mustang transmissions 1974-90

	Code
Four-speed overdrive (SROD)	4
Five-speed manual	5
Five-speed manual overdrive (RAP)	5

Mustang transmissions 1974-2000

	Code
Four-speed manual (Borg-Warner)	6
Four-speed overdrive (RUG)	7
Four-speed (ET)	7
Three-speed automatic *C3*	V
Three-speed automatic *C4*	W
Three-speed automatic *C5*	C
AOD (automatic overdrive)	T
C-6 Police automatic	Z

Mustang specifications

	1965-66	1967-68	1969-70	1871-73	1974-78	1979-93	1993-95
Wheelbase, in.	108	108	108	109	96.2	100.4	101.3
Track, frt./rear, in.	56/56	58.1/58.1	58.5/58.5	61.5/61.5	55.6/55.8	56.6/57	60.6/59.1
Width, in.	68.2	70.9	71.8	74.1	70.2	69.1	71.8
Height, in.	51	51.8	50.3	50.1	50.3	51.9	52.9
Length, in.	181.6	183.6	187.4	189.51[1]	175	179.1	181.5
Curb weight, lb.	2,860 (289)	2,980 (302)	3,625 (428CJ)	3,560 (351CJ)	3,290 (302)	2,861 (140)[2]	3,065 (3,276 GT)
Wt. dist. % f/r	53/47 (280)	56/44 (302)	59/41 (428CJ)	56.5/43.5 (351)	59/41	57/43[3]	57/43

[1] 193.8 in 1973
[2] 3075 with 302
[3] 59/41 with 302

	1996–1998	1999–2000
Wheelbase, in.	101.3	101.3
Track frt./rear, in.	60.0/58.7	59.9/59.9
Width, in.	71.8	73.1
Height, in.	53.4	53.2
Length, in.	181.5	183.5
Curb weight, lb.	3466	3430 (Conv. 3560)
Wt, dist % f/r	57/43	55.5/44.5

Rear axle codes — 1965–2000

Axle	1965	1966	1967	1968	1969	1970	1971	1972	1973	1974	1975-76
2.35					F						
2.47											
2.50				0							
2.73											
2.75			8(H)	1(A)	2(K)	2(K)	2(K)	2(K)	2(K)		
2.79				2(B)	3	3	3	3	3		
2.80	6(F)	6(F)	6(F)	3(C)	4(M)	4(M)	4(M)				
2.83	2(B)	2(B)	2(B)	4(D)	5	5					
3.00	1(A)	1(A)	1(A)	5(E)	6(O)	6	6	6	6(O)		6(O)
3.07						B	B				
3.08					C(U)	C					
3.10					7						
3.18											
3.20	3(C)	3(C)	3(C)	6(F)		8					
3.25	4(D)	4(D)	4(D)	7(G)	9(R)	9(R)		9(R)			
3.27											
3.40											7
3.45											
3.50	5(E)	5(E)	5(E)	8(H)	A(S)	A(S)	9(R)	A(S)	9(R)		
3.55											
3.73									A(S)	G(X)	G(X)
3.89	8(H)	8(H)					A(S)				
3.91					(V)	(V)		(V)			
4.11	9(I)	9(I)	9(I)				(V)				
4.30					(W)	(W)	(Y)				

Letter in parentheses indicates locking differential

Rear axle codes—1965-199?

Axle	1977-78	1979	1980	1981	1982	1983	1984	1985	1986	1987	1988	1989	1990	1991	1992	1993	1994	1995–2000
2.35																		
2.47		B	B(C)	B(C)	B(C)													
2.50																		
2.73			8	8(M)	8(M)	8(M)	8(M)	8(M)	8(M)	8(M)	8(M)	8(M)	8(M)	8(M)	8(M)	8(M)	8(M)	8(M)
2.75																		
2.79	3																	
2.80																		
2.83																		
3.00																		
3.07																		
3.08		Y(Z)	Y(Z)	Y(Z)	Y(Z)	Y(Z)	Y(Z)	Y(Z)	Y(Z)	Y(Z)	Y(Z)	Y(Z)	Y(Z)	Y(Z)	Y(Z)	Y(Z)	Y(Z)	Y(Z)
3.10																		
3.18	4																	
3.20																		
3.25																		
3.27						5(E)	5(E)	5(E)	5(E)	5(E)	5(E)	5(E)	5(E)	5(E)	5(E)	5(E)	5(E)	5(E)
3.40																		
3.45		F(R)	F(R)	F(R)	F(R)	F(R)	F(R)	F(R)	F(R)	F(R)	F(R)	F(R)	F(R)					
3.50																		
3.55								(W)	(W)									
3.73																		
3.89																		
3.91																		
4.11																		
4.30																		

Letter in parenthesis indicates locking differential